EVE AND THE APES

ALSO BY EMILY HAHN

EVE AND THE APES

Emily Hahn

WEIDENFELD & NICOLSON
New York

Published by Weidenfeld & Nicolson, New York
A Division of Wheatland Corporation
10 East 53rd Street
New York, NY 10022

Published in Canada by General Publishing Company, Ltd.

LIBRARY OF CONGRESS CATALOGING-IN-PUBLICATION DATA

Hahn, Emily, 1905–
    Eve and the apes / by Emily Hahn. —1st ed.
        p.   cm.
    ISBN 1 555 84172 4
    1. Primates as pets.   2. Primates.   I. Title.
    SF459.P7H34   1988
    636'.988—dc19   87-34008

Due to limitations of space, permissions and credits appear on page 180.

Manufactured in the United States of America
Designed by Irving Perkins Associates, Inc.
First Edition
10 9 8 7 6 5 4 3 2 1

for Robert and Roberta Yerkes
with great affection

ACKNOWLEDGMENTS

IN WRITING THIS BOOK, the author called upon many resources, not the least of which, of course, is the goodwill of the women written about. But several books should be acknowledged particularly: *My Life in a Man-made Jungle!* and *My Friends, the Apes*, by Belle Benchley; *Almost Human*, by Robert Mearns Yerkes; *Toto and I*, by Maria Hoyt; *The Education of Koko*, by Francine G. Patterson and Eugene Linden; *Silent Partners*, by Eugene Linden; *The Chimps of Mount Asserik*, by Stella Brewer; *Apes on Stage* and *Animals Are My Hobby*, by Gertrude Davies Lintz; *Gorilla Show*, by Mae Noell; and *Orang-Utan*, by Barbara Harrisson. A complete bibliography can be found at book's end.

TABLE OF CONTENTS

Eve and the Apes

$\mathcal{P}$ROLOGUE

"NOW THAT'S THE KIND OF THING I can't sympathize with," said the man on the ship. We were aiming, as I remember, for India, not that that had anything to do with it except that on a long voyage one tends to fill in the time with prattle, and I had displayed enthusiasm for the gibbons of Thailand. Why gibbons, why Thailand? Well, why not? It used to be like that on shipboard. As I prattled, my mind went back to the happy days before the war in Shanghai when I actually kept gibbons. But as I talked, that man's face showed—could it be?—disgust. I never talked to him again.

I have traveled a lot, as anybody who reads this book will gather. It started late, this indulgence—at least I consider twenty-two late—but I have made up for it since, and there are few countries I haven't visited, at least this side of the iron curtain. How did I do it? It was easy, once I learned the trick of concentration, saving up for this and that journey, and it became easier still when I became a journalist. Naturally I tried to concentrate on what I liked best, animals, and of the nonhuman animals my favorites have always been the primates. Sometimes it was a direct bit of research, as when I went around the world inspecting zoos, and sometimes it was something I smuggled in on the side, as when, during a nine-year stay in Shanghai, I simply bought all the primates I could find—Indian rhesus macaques, imported by the Chinese for God knows what purpose, and gibbons from the south. In what used to be called the Belgian Congo, monkeys and chimpanzees were all over the place. I was happy there.

All of which should go to show that I am interested in apes; indeed, I am enthralled by them. I met in New York my first nonzoo chimpanzee, a young female that had come to America with her owner on a visit. I followed up the acquaintance on the Continent, where it was necessary for me to take care of her for a few days, and then I picked up our friendship again in the Belgian Congo. I will try to recall just what it was about

3

Chimpo that reached out and caught me. That is, if I know, myself. First, it was her eyes—light brown, wrinkled around the edges, questioning, intelligent, and imploring. Next, I guess it was because she was so much like a human child. Yet at that time I didn't care all that much for human children. Anyway, I was fascinated. Caught. I realize that this fascination doesn't always carry over to other people. Some humans have expressive eyes of their own, which tend to glaze over when I get started on the subject of apes. Others are disgusted.

Obviously I wasn't. It was not only great apes I loved; I acquired and looked after a number of primates, smaller monkeys and especially baboons. Baboons are not everyone's cup of tea, but they are decidedly mine. I had nine of them of different ages, and they made a lively troop. They lived free; after all, we were in the middle of the Congo forest, and there was no reason they wouldn't. Once a baboon has decided he is tame, he sticks around, as did my nine. They had the run of the place; if I hadn't let them take it, they would have anyway. The first time I saw human disgust for nonhuman primates was when I was entertaining the village headman, or capita, at tea, and my youngest baboon came rollicking in and pulled at his skirt (the capita was an Arabized African, a Muslim). I knew that the Muslims there didn't like dogs, and of course, they hated the wild pigs that sometimes rampaged in the woods; but I hadn't realized that their fastidiousness extended to apes and monkeys.

Later, in China, I missed the baboons and was glad to find their distant cousins, rhesus macaques, available. Dr. Mary Pearl, an authority on the species, once told me severely that I was mistaken, that what I had there in Shanghai could not have been the rhesus but the Japanese macaque. However, I have persevered in researching. They were rhesus. Trade between China and India took care of that. One doesn't give macaques the freedom of one's house, not if one has any sense whatever, so they lived in a cage in my front yard.

It was by happy chance that I first saw the gibbon Mr. Mills in a window of a shop down near the Whangpoo River. He was tied by the neck with a too-short string, and we looked at each other, and I dashed in and bought him. I had never seen a gibbon. Indeed, I had no idea what Mr. Mills was; for all I knew he could have been a possum, but he was a gibbon. Mr. Mills *did* live in a house—no nonsense about the front yard for him. So there we were, the gibbon, the macaques, and I, and—unfortunately—the lady next door, who hated us all. I don't think the macaques ever got into her garden, but Mr. Mills did, in his insouciant way, wander over to eat a flower now and then, especially as her daughters lured him with food. How could anyone, I still ask myself, object to a gibbon with his black pansy-shaped face, his enormous round eyes, and his incredible grace? But the woman next door—an Englishwoman, at that—did object, and she went to the police to complain about me. I am happy to say that the police saw things my way, and my household continued to be, more or less, undisturbed. There was an agitated neighbor—Chinese, this time—who threatened to shoot Mr. Mills after the gibbon had raided his kitchen, but the SPCA fixed *him* pretty quickly.

I can see that nonhuman primates might seem to some people like rude mockeries of ourselves, an affront to our humanity, but I've never felt affronted by them. Fortunately not everybody has difficulty with primates. Nobody objected when I kept a monkey or two in my New York house, or even when I had, for a few sad days, a dying chimpanzee that I was taking care of for a friend. But New York is not a good place for apes and monkeys, and England, where I sometimes live, is out of the question. Both places have difficult climates, except perhaps for golden monkeys from western China, and they are very rare; I have never owned one. Yet even in indulgent New York one runs into pockets of resistance, like the veterinarian I met the other day in that city. Surely you would think, an animal doctor . . .

"Monkeys? I hate 'em," she said roundly.

"Well, maybe, but the great apes—"

"I hate 'em all. I won't take care of them in my hospital. They stink. They're into everything. They—" she paused and drew breath. "They look *just like children*."

Our species is rich in foster mothers who willingly take care of children not their own, but I don't think that is the same impulse that motivated the women I have written about in this book. Maternal impulse plus the fascination of the unknown? Possibly. Probably. Is caring for apes the same urge that has so many teenage girls mucking out livery stables and grooming ponies in order to be close to the horses they love? One cannot attend a point-to-point without being impressed by the number of girls competing, compared with the scanty representation of boys. Many a riding establishment could not carry on if it were not for those devoted young women who are willing to work for small or no reward beyond the occasional free ride. And I get a lot of letters from girls who would like to break into the zoo world so they can take care of animals. This may be because we live in a civilization where conservation is the cry as urbanization closes in on us— but where are the boys in all this?

It would be easy but not satisfying to dismiss the question "why?" about these ape keepers. We could say, "Sheer maternal instinct," and dismiss the subject, but would that explanation quite fill the bill? I don't think so. Many of the women I have discussed are not in the least frustrated as mothers. They have children as well as apes. There are some others I haven't singled out, such as Mrs. Reynolds, wife of a primatologist, who kept a young chimpanzee with her son of the same age, making notes on their parallel development until it became obvious that the experiment wasn't completely satisfactory: the son was very slow learning to talk and began barking for his food.

No, maternal instinct is not quite the answer. Perhaps these women (like many other humans) find it easier to get along with nonhuman animals than with their own kind. How many times have you heard people declare that they prefer animals to other people? "The more I see of people, the more I prefer dogs." It's a very common attitude. However, there is another possible explanation. The women I have written about, the adopters of apes—most of them lived at a time when it was taken for granted that their sex is the nurturing one. They were the mothers, and of course, people thought they felt maternal toward these young almost-humans. Today, however, we are not quite so set in our ideas, and we give some men full credit for being just as sympathetic and gentle with animals as are women. But this is a very recent development. Twentieth-century women are emerging from earlier stereotypes, but in that one stereotype they have until very lately been firmly fixed—the nurturing sex. That a woman should be good at rearing an exotic creature like an ape was accepted as natural, even as a safe means of self-expression.

Male or female, it's clear that few people are indifferent to nonhuman primates. Apes and monkeys arouse feelings in most humans—emotions, it may be, of fear, attraction, disgust, or simply curiosity. With me it is fascination, and I know I am not alone in this reaction because whenever I have occasion to go out in public carrying a monkey or an ape, people who see us register something, attraction or, perhaps, irritation—the kind of irritation, one might say, aroused by modern painting or sculpture. . . . "Is the artist, or the monkey, making fun of me?" But whether it be attraction or anger, it is *something*; people register.

As time went on and I stopped keeping apes, an urban existence making it impossible to go on living with them, it became increasingly evident that the majority of men do not share my passion for primates. I say men rather than women, and I say it advisedly, because my pets' best friends have always been female. I know that the odd case does not prove my point, but there is no law against citing outstanding examples, such as the time in Shanghai when I was entertaining the writer John Gunther and his wife, Frances, at lunch. I owned my first gibbon then. During cocktails he came into the room, wearing, as he always did, a diaper, what the English call a nappy, a costume that seemed to be then, as it still does, an eminently sensible thing to put on a domesticated ape, though for some reason the sight of it outrages some people. It didn't outrage Frances Gunther, but she didn't like the gibbon at all, not at first anyway. I noticed that she shrank back whenever he came close to her. After half an hour or so, realizing that he needed to have his nappy changed, I carried him upstairs to my bedroom, and Frances came along, saying she wanted to watch the process. Critically she did so, at first from a distance, but suddenly she said impatiently, shouldering me out of the way and grabbing the safety pin, "Here, give me that. You're doing it all wrong," and she deftly changed the nappy. John, on the other hand, pretended the gibbon did not exist when we went back to the party. This was not a good idea, as gibbons cannot bear to be ignored. Mine concentrated on him from then on.

The history of animal husbandry is full of examples of women who have become fascinated with primates and who, in their turn, seem to fascinate the primates they take care of. There are so many of these ladies that the list will probably surprise you with its length, and even so, it is incomplete. Nevertheless, the account of such women must be preceded by the masculine Dr. Robert Mearns Yerkes, psychobiologist and father of primatology in the United States. His wife, Ada Watterson Yerkes, encouraged him when, as a young psychology professor in 1923, he spent nearly all of his little family's funds, two thousand dollars, on a pair of chimpanzees. He had long wanted specimens of anthropoids for his studies, and Mrs. Yerkes understood his desire and shared it. Four years later the couple produced their classic work *The Great Apes*, and it is doubtful if they could have done so without these two little animals, Chim and Panzee, in the background. I shall recount Mrs. Yerkes's story later; for the moment I shall name some other women who were to help Dr. Yerkes's research: Madam Rosalía Abreu, the heroine of his book *Almost Human*; Blanche (Mrs. William S.) Learned, a musician and musicologist who wrote of chimpanzee vocalization; Cathy Hayes of the Orange Park primate laboratory; Nadazhda Ladygin-Kohts, wife of Aleksandr E. Kohts, who was curator of the Darwinian Museum in Moscow. These are only a few of the women who worked with primates. The closer to the present we come in the history of primatology, the more women's names crop up, until we come at last to those the public considers the Big Three: Jane Goodall with her chimpanzees, the recently murdered Dian Fossey and the mountain gorilla, and Biruté Galdikas, whose specialty is the orangutan. But this is not an exclusive list. Today we must also think of Jo Fritz, who, with her husband, Paul, maintains a breeding colony of chimps in Arizona, and the late Dr. Gertrude van Wagenen, with her carefully maintained Yale group of rhesus macaques, studied for knowledge on reproduction, who has probably done more for birth control (against her own conscience—she was educated in the Roman Catholic faith) than any contemporary, and Sue Savage Rumbaugh, who, after a spell with the psychology professor Dr. William Lemmon's chimpanzee colony at the University of Oklahoma in Norman, has worked for the past few years with Bonobo chimps at the Yerkes Primate Center at Emory College, Atlanta, Georgia. And there are many more. Clearly women are adept at animal maintenance and always have been. Since the trend seems to have started more or less all at once in the twenties, I can't think of a better person with which to initiate the subject than Belle J. Benchley of the San Diego Zoo.

1. BELLE BENCHLEY: ZOO DIRECTOR

WAS MRS. BENCHLEY actually the zoo director by title? Not imme-diately, at any rate, nor did she claim to be an authority on the subject. In her three books she explains how it came about that she achieved that eminent position. In her native California her first job was teaching at a little Indian school in the country, before the zoo itself was created in 1916. It got its start through the devoted efforts of Dr. Harry Wegeforth and some of his cronies, leading spirits all of them in the then small city of San Diego, where a fair, the Panama-California Exposition, had just wound to a close. Here and there near the deserted fairground, washed up like driftwood, were cages holding miserable animals whose owners had run out on them because it cost too much to go on maintaining them now that there were no more visitors to the fair. Dr. Wegeforth, a prosperous physician, was affronted by the situation. He was fond of animals (one of his relations kept wild animals as a hobby), and he resolved to do some-thing about these orphans. Why shouldn't San Diego support a zoo? It was a growing town with a delicious climate. Here were a few animals for a nucleus; all they needed was space and a supporting fund. Dr. Harry, as his friends called him, set to work vigorously to scrape together site and money. He got a tract of land at the edge of Balboa Park—sandy, deserted, and no good to anyone else, as he often pointed out—and then he forced many of the more prosperous citizens to contribute to the project.

For her part, Mrs. Benchley had given up teaching for marriage and motherhood; by 1925 she had a son of seventeen but no husband and no job. Very much on her own, she was glad to accept the job of bookkeeper at the zoo when it was offered to her by Dr. Harry. She knew nothing of bookkeeping, as she wrote in her first book of memoirs, *My Life in a Man-Made Jungle!*, and less about animals, but she was more than willing to find out about both subjects.

"Even now," she wrote in *Jungle*, "I fail to recognize the steps leading to my advancement to the headship of the zoo staff." It might have been, she thought, because the zoo was still very poorly organized. At any rate she had been embarked on her new job only an hour when the telephone rang and she heard a man say that he and a friend wanted to settle a bet and were depending on her answer to do so. "How long is the tail of a hippopotamus?" was their question. What Mrs. Benchley replied is not on record; but it is certain that she asked for time to go and look it up, and surely she gave them an answer later on. People continued to ask questions. During the next few days there were more telephone calls. Somebody's canary was ailing; what was the matter with it, and what should be done? Mrs. Benchley went and asked the local vet for advice. Somebody else wanted to know the gestation period for elephants; Mrs. Benchley found out and no doubt was never to forget it afterward. A snake was sketchily described: What species was it; was it dangerous? Mrs. Benchley consulted a herpetologist.

Did she ultimately have to hire another bookkeeper? Probably, although not for some time. "At the end of a year and seven months," she tells us, "I was advanced to my present position."

They were pleasantly simple days. Today the director of an up-to-date zoo must face a battery of examiners and undergo questions that test the knowledge of the most learned and experienced of candidates, but Mrs. Benchley earned her position in her own way. She worked very hard and took on great responsibility. Not only did she read everything she could find on the various species of animals in her care, but she went out and gave talks about the zoo, for the good of the establishment: talks in clubs, schools, lecture halls, anywhere or nearly anywhere that she was invited to talk. In 1943 she reported that during the year just past she had given 150 lectures, and requests for more kept coming. The San Diego Zoo, like San Diego itself, was growing fast in size and repute. Belle Benchley's friends no longer joked about her appointment; it was no joke, but a serious job, and she was doing remarkably well at it. More important in the last analysis was that she had discovered within herself a talent, hitherto unsuspected, for dealing with animals. She announced that the secret of a good zoo director is based upon knowledge of the animal world, based on general study and continuous intelligent observation of his or her collection. His or her interest must be so great as to amount to a curiosity which can never be satisfied.

And a zoo director must be self-confident to an extreme degree. Witness the time, early in her career, when a pair of the zoo's baboons escaped through a door which had been carelessly left open. Like many other wild animals, baboons can be dangerous when they feel threatened, and if everybody around them began chasing this pair, they would feel threatened. Mrs. Benchley felt that at all costs a hue and cry must be prevented. Fortunately she was fairly well acquainted with these animals, having made it a part of her job to circulate through the collection as often as her duties permitted, giving some of the animals little tidbits. She felt she knew the male baboon pretty well, and so, evidently, did he, for when she walked up to him and offered to take his hand, he permitted her to do so. Quite calmly he trotted along, the keeper holding his other hand, with the female following at their heels. Together they walked back to the baboon cage, and the animals went in. It was that easy. As she was to learn, most animals do wish to return to their cages, preferring a familiar place to the great unknown outside.

The publicity that attached to Belle Benchley's position—"woman direc-tor" and so on—disturbed her a little at the beginning, but soon she was used to it. She even became reconciled, as she wrote in *Jungle,* "to seeing myself in print in ungainly smocks such as no stout woman should ever wear and with my hair streaming in what one reporter said was 'my particular style,'" because, as she cannily remarked, a good news story is always news, and news was (and still is) the lifeblood of the zoo. It is true that she was, as she confessed, a plump woman; in fact, she looked like a good-tempered granny or cook rather than a zoo director, but after all, this happy, comfortable appearance was thoroughly compatible with her atti-tude toward her charges. Her relations with the great apes, the animals that probably attracted the most publicity for the San Diego Zoo, were decidedly grandmotherly, which can be seen clearly in her book *My Friends, the Apes.*

Of course, life at the zoo was not one unbroken success story. Mrs. Benchley had her troubles, among which was the fact that some male Californians resented a woman's being director. We get a hint of this attitude—well, more than a hint actually—in her account in *Jungle* of a tragedy that overtook the bear community. San Diego had a good, flourish-ing colony of grizzly bears. One of the females, Babe, gave birth to cubs several times and proved herself an exemplary mother. Then one day, when the zoo was full of visitors—a fine day, as it so often is in San Diego—a few of the grizzlies found an open gate and got out. The tranquilizer gun had not yet been invented, and one of the police force, fearing for the safety of the public, shot and killed Babe. Mrs. Benchley wrote in *Jungle*:

> As always, much criticism was heaped upon the zoo management, and the police officer who shot the bear, and one priceless card came to me which told me that it was all because I was a woman and incompetent; that if I had had a man in charge of the zoo, the bear would have been lassoed and dragged back into the grotto. Well, of course, the head keeper was a man in complete charge of the animal men; the keeper of the bears was also a man, and so was the police officer who fired the fatal shot. Moreover, the one case on record where the lasso was ever employed to capture an animal in such circumstances resulted in the death of the bear in horrible agony because, in his fury and excitement at being lassoed, he pulled so hard that he literally tore himself in two before they could release the lasso. . . . We feel that we are fortunate that in San Diego a severe injury, either to a keeper or a visitor, has never resulted from an escaped animal or from the entrance of anyone into an unprotected cage.

Mrs. Benchley's special fondness for apes seems to have started with gibbons. Like a lot of other people, she had never particularly noticed the species before she joined the zoo, and for some time after that she thought, erroneously, that they were monkeys; but soon she learned better. Conscientiously, as usual, she started out to get better acquainted with these graceful creatures. She wrote in My *Friends*: "Anthropoid apes were something of which I had heard, of course, and in some dim way I had connected them with Darwin's theory of the Descent of Man, but that is all. I was instantly attracted to these thickly furred monkeys with their small black faces. . . ."

Soon she found herself, in her first exploratory peregrinations, going back over and over to watch the gibbons. It was a long time before she discovered that they were not monkeys or noticed that they had no tails; she had to find out such details by herself. At first the zoo had only a pair of gibbons, and Mrs. Benchley observed that the female, Gibby, was a bully, constantly taking the best of any meal for herself and leaving the rest for her mate, Billy, who seemed to take this treatment philosophically. It didn't happen only at their regular mealtimes. In those happy-go-lucky days zoo visitors were permitted to feed the animals, and Gibby did the same thing with such extra handouts, grabbing the lion's share, while Billy had to be content with what she discarded. Belle Benchley worried about Billy's welfare and talked the matter over with the head keeper, an experienced man named Henry Newmayer. Yes, he said, it wasn't a good situation. He agreed with Mrs. Benchley's suggestion that Gibby be deprived of her wicked long canine teeth, and the sooner the better. But when he tried to catch Gibby, there was trouble. Billy leaped to her defense, and it was *his* canine teeth that caught Henry in the neck, fiercely enough to draw blood. The project was postponed to a later date while Henry's wounds were dressed.

In the course of time the San Diego Zoo acquired two larger gibbons of the species known as siamang, along with several more specimens of smaller ones. Soon the staff was hearing the awe-inspiring concert given by gibbons when they sing at dawn. Siamangs are especially vocal. They have sacs in their throats that swell and throb as the apes hoot, and the noise they make is incredibly loud as well as musical; once one has heard it, it is not easily forgotten. Their song reminded Mrs. Benchley of something she could not quite place for a while, but at last she remembered a "large and talented Indian" who used to play for the dances that were held in the neighborhood of the school. This Indian was a one-man orchestra. He carried a mouth organ, a drum, and even a stringed instrument like a guitar or a banjo. With all of these instruments going at once, she wrote in *My Friends*, he produced "something of the medley of sounds not unlike the combination of many of the tones produced by the hard-working gibbons." One day the singer Nelson Eddy visited the zoo and was very interested in the song of the siamangs. He explained to Belle Benchley that in spite of the resonance of the ape's voice, which sounded like several noises at once, she (the gibbon) actually produced only one sound, which she first emitted through her mouth, then forced through the stretched skin of the laryngeal pouch.

"This had a tendency to muffle and deepen the sound and at the same time to give it a resonance associated with such instruments as the drum." Nelson Eddy boomed at the siamang in her own language; she answered; he replied to her answer; for some time man and ape sang together. Soon the other gibbons joined in.

By this time the reader will probably have realized that Mrs. Benchley never did acquire—if, indeed, she wished to, as is doubtful—a hard-boiled scientific attitude toward her animals. She expected from them a certain moral commitment, at least from the great apes; in today's jargon, she projected her own code of behavior on the animals. We see it over and over. When Gibby snatched food out of Billy's grasp, Belle Benchley was shocked by the female's injustice. When Billy attacked the keeper because the keeper was trying to catch Gibby, Mrs. Benchley admired his chivalry. She never stopped identifying with the animals. She thought of the other species as akin to herself, an attitude that probably did no harm whatever to the San Diego Zoo. She was never guilty of detachment.

She showed the same trait in her dealings with the orangutan. Until some time after she had joined the staff, the zoo had no specimens of this red-haired ape from Indonesia, the "man of the forest," as the name is translated. This was true until March 1928, when an animal dealer loosely connected with San Diego—most dealers worked on a free-lance basis—arrived with a few animals he had collected in Southeast Asia, at his base in Singapore. He had with him two little Malayan bears, a gray gibbon, a few birds, and two very young orangs. The smaller of these last, a male, could not have been more than a year old and was clinging tightly, with both arms around her body, to the other orang, a female about a year older. "She held the baby against her side astride her hip much as I have seen older children lug the heavy baby," wrote the enchanted Mrs. Benchley in My Friends. "She was not so shy but appeared just as innocent. They were both covered with an old quilt or robe. . . ."

The zoo personnel immediately named the red-haired girl Maggie, and of course, the little male was dubbed Jiggs. It was sad when they had to part from the collector, a young man who had cooked porridge for them every day aboard the ship that brought them to California; they held out their arms to him and cried when he departed, and he, too, nearly cried. He had taught them to eat their porridge out of bowls, with spoons. However, when Mrs. Benchley held out her arms, they came directly to her, too. "I was prouder than the occasion warranted," she confessed. ". . . I was to learn that aboard ship they had been everybody's pet and had lost all fear of strangers. I was just one more kind and safe human being to them."

Each cage in the primate section of the zoo was constructed with a downstairs, or exhibition, cage and an upstairs cage walled off from the public, the animal's bedchamber. In preparation for Maggie and Jiggs, clean hay and gunnysacks had been placed in their bedchamber, but when

Maggie saw the arrangement, she set to work immediately to change it. Jiggs fell asleep upstairs, but she threw the hay and sacks down to the lower cage and began to arrange a bed. She did this so carefully that Henry Newmayer went to fetch Mrs. Benchley to see what was going on. First Maggie pushed the hay together until it was neatly shaped into an oblong mattress. Then she spread a gunnysack over it, and after that a second sack, and after that a third. She was hard to satisfy, going over and over to make sure the whole thing didn't get messed up. At last she went upstairs, grabbed the sleepy Jiggs, brought him down, and planted him in the middle of the bed. With the fourth and last sack clutched in her hand (or foot) she joined him, pulled the sack over both of them, and seemed about to sleep. But she still was not quite content; after a minute or so she sat up and looked the terrain over, got out, and tried to make the bed once more. Finally she grew tired of it all and lay down anyhow, the sack huddled around her in the middle of the hay. Both orangs were soon asleep.

So far so good, but Mrs. Benchley, like Henry, was soon to discover certain difficulties in the care of orangs. They are incredibly strong, and when they want to investigate their surroundings, little stands in their way. Maggie immediately took apart the small wooden chairs the keepers supplied for the porridge party, which had been introduced as a display feature; two much sturdier chairs had to be obtained. In time a lot of other changes had to be made in the cage. The orangs' prying fingers could undo bolts as if they were hairpins. At least Maggie's fingers could; little Jiggs did not survive very long, dying at the end of the summer. The autopsy showed what had killed him: a wad of chewing gum, complete with paper, which he had been too weak to excrete, one more case for forbidding the public's feeding of animals. (It is amazing that even good zoos like San Diego's took so long to outlaw the practice.) Left alone, Maggie turned to the keeper Henry for solace, and he willingly accepted the role of special friend, taking her home with him at night and generally letting her do anything she wanted. By the time the zoo acquired another companion of her own species to keep her company, a much older female who had outgrown her usefulness as a Hollywood actress, Maggie was a spoiled little ape. The newcomer, named, inevitably, Big Jiggs, bullied her to some extent; but although ill tempered with everyone else, she never actually bit Maggie or otherwise mistreated her, and Mrs. Benchley appreciated the fact that these things were important. But here, too, she judged the apes from a moral standpoint. She felt that Maggie was a *good* girl, she took such loving care of Little Jiggs. She was also a careful housewife, who kept her cage remarkably clean.

To be sure, in addition to these virtues, she had some more questionable talents, but her efforts were praiseworthy, even if awkward. For example, she had great ability for building nests in acacia trees if she could get at the trees. Most zoologists know now that orangutans in the wild show plenty of mechanical ability, but Mrs. Benchley, with other zoo people, had to find out for herself. Said the director with rueful pride in *Jungle*:

> She has never permitted herself to be held in her sleeping room by any mechanical device such as ratchet or drop bars and has been known to exert such strength and cunning that she has slipped padlocked bars out of their grooves far enough to get a door open and get out. All of us agree with the keeper who said: "Maggie! Oh well, I expect to come in any morning and find Moore locked up and Maggie out swabbing the cage."

As Maggie matured, naturally she grew even stronger. Mrs. Benchley told, in *Jungle*, how she idly took a single steel link from a strong metal chain, twisted it, and flattened it into an almost straight piece of metal. There was no obvious reason for this exercise; it seems to have been accomplished merely because, like Everest, the chain was there. "Anyone would have doubted that it could be done with cold metal, even with a vise," said Belle Benchley. And when workmen put up on the orang cage a bronze plaque honoring the donor who had provided funds to build it, though the bolts holding the sign were three inches from the cage wire, Maggie managed to remove them all so quickly that the sign didn't stay in place more than a few minutes. Of course, she was not the only destroyer; she was aided and abetted by Jiggs. In Mrs. Benchley's estimation Jiggs was the "smartest" ape, regardless of species, ever known to the director. She wasn't lovable, but she was decidedly smart. One day, for instance, she somehow got hold of a large pipe wrench—probably stealing it from workmen mending the cage—and in a short time she learned how to manipulate the thing. She was having a great time with it when she inadvertently hit Maggie in the head, nearly braining her with the heavy tool. Mrs. Benchley had observed the incident and discussed it with Henry, who agreed that the orangs would be better off without the wrench. But how to get it away?

Jiggs fully appreciated the value of her prize and was canny enough to take it everywhere with her. The humans tempted the apes with a selection of food in the sleeping cage, but as they rushed in to eat, Jiggs brought the wrench with her. Then Henry tried offering her a number of other articles in exchange. No bargain: Jiggs kept her hand on the wrench. Baffled, Henry shut the door to the bedchamber and went away to ponder his next step—and in so doing settled the matter, for Jiggs, behind the closed door, decided that she was being punished and violently reacted, pounding on the door to be let out. When Henry came back, she handed him the wrench without any more palaver.

Of course, Mrs. Benchley applied the adjective "smart" to animals that adapted themselves in this manner to the human viewpoint, and it was natural that anthropoid apes should be "smarter" than other nonhuman animals. After all, they *are* anthropoids.

Big Jiggs went off to another zoo, and after a little more time the authorities managed to replace her with a strong young male orang named Mike. It was a fortunate exchange, a happy marriage. But there was one more orang in the zoo that the director felt she must mention—Katjeung, whose arrival was heralded by Dr. Harry when he said to Mrs. Benchley, "Old girl, Karl Koch [one of the zoo's collectors] is bringing you the biggest orang you ever saw!"

She did not take this remark too seriously, as one of her standing jokes with Wegeforth was that his glasses always magnified the animals he bought. But this time he had not exaggerated. The orang weighed two hundred pounds and had been confined, perforce, in a cage weighing fourteen hundred. Mrs. Benchley wrote in *Jungle*:

> The crate was made of steel rods set in heavy steel plates and lined with teakwood or mahogany. Outside this wood was sheet-metal, but the inner walls and wooden shelves had been splintered by the teeth and hands of the orang during the forty days he spent in transit. He had bent and twisted the steel bars in the front until Karl, fearing he might reach out and injure someone, had wound a heavy chain, padlocked at the ends, back and forth across the front. A large sign had been attached to the case, reading, DANGER! KEEP AWAY!

Katjeung was so large that each of his fingers was as thick as Belle Benchley's wrists. The construction manager took one look at him and went to reinforce the cage where the giant ape was to live. Once he was installed—having whirled across the cage in the strange manner of orangutans—he placed himself on a lower ledge in his quarters from which he could look straight into the cage of Maggie and Mike and grind his teeth at Mike. Orangs, when they are stirred by emotion, always grind their teeth, this being the only sign they give of their turmoil. Mike ground his right back. In fact, the two males did nothing for days but grind and gnash their teeth at each other, until at last Mrs. Benchley felt she must take a hand, and here it was that her special talents came in useful, Maggie being no good at all as a peacemaker, since every time she came over to investigate Katjeung at closer quarters Mike became jealous, naturally egging her on. As for Katjeung, he was so worked up that he couldn't eat. No, it was up to Belle Benchley.

It was not a matter of being fearless. She knew all about the orangutan's characteristics—slyness, cunning, enormous strength, and unexpected swiftness—and for the first time in her career she felt afraid. She had always simply loved apes, but Katjeung was different. Carrying a lot of food in her hands, she came close to the cage, noting that his long, dark, hairy, curled fingers were at least ten inches long. If her hands came too close to the wire, she was sure he would grab them. Still, something must be done. She walked between the cages, stopping a little to obscure Katjeung's view of Mike. To see the other ape, he would have to move, but he didn't. He looked straight into Mrs. Benchley's eyes for the first time since his arrival. Now she began to talk to him, saying his name in a quiet voice and very slowly moving closer. Gradually Katjeung stopped grinding his teeth, even though behind her back she could hear Mike still grinding, still threatening. After a few moments of this concentration she thought she saw a new expression in Katjeung's eyes, a kind of curiosity, as if he were studying her. Very slowly she lifted her hands, full of fruit. She heard from Katjeung a new sound, a kind of outblown breath that was almost a voice, as she showed him the halves of a large peach and held one half of it against the cage's wire.

"Slowly, suspiciously, and with no appearance of eagerness, he moved that tub-sized face forward," she wrote in *Jungle*, "and, extending his lips, took the peach in them. He manipulated it out of the fuzzy skin, still with his lips, and slowly ate. He took the second half in his fingers . . . and went back to his serious contemplation of my face. He showed no interest in the other food. . . ."

18

Slowly their friendship made progress from day to day. After a while he often accepted food from her hands and moved around to face her as she walked by the cage. One day, as she came past, he abandoned staring at Mike, climbed down off the shelf, and followed her as far as he could inside to the end of the cage as she walked on to the chimpanzees. She felt a thrill of pride in this, but the conquest was complicated because Katjeung now resented her paying attention to others, even members of the public. Then he fell ill of an infection in his great cheek pouch, and although in the end he was treated and cured, he was a long time away in the hospital. During his absence the zookeepers needed his cage and put in two newly arrived young orangs, so Katjeung's convalescence, a fretful one, had to be undergone in new quarters where the walls had been hastily reinforced. What with one thing and another it was a definite relief when he was exchanged and sent away to an eastern zoo, which already had a cage strong enough to hold him. Though, as we see, he had, in spite of appearances, been reasonably amiable with Belle Benchley and even with the young woman who tended him during his illness, the ladies agreed that an orangutan of that size and tetchy disposition was a great, worrying responsibility. Yes, it was a relief when Katjeung went away.

On one occasion Mrs. Benchley, along with a number of other interested colleagues, was able to visit Cuba and see the primate colony maintained by Madam Abreu in Havana. There she immediately fell in love with a baby orang that was living at the Quinta Palatino with his mother. No orang had as yet bred in San Diego, and Mrs. Benchley was ecstatic over this infant. She was moved to heights of gushiness—there is no other word for it—that she seldom permitted herself. She could not believe that the great, rotund, coarse-skinned mother could have produced such a perfect baby. As she approached the cage with her hostess, the face of the infant screwed into an entrancing grin.

He had recognized Madam Abreau! [she wrote in *Jungle*, misspelling Madam's name]. . . . from that second I have wanted to hold and play with and keep close to me a baby orang. I know they are difficult to raise, that they become mean, headstrong and destructive, and that they are mischievous and always into things, but all this is entirely offset by their thousands of cute expressions. No other baby animal has anything like the innocence, the confidence, and the complete dependence upon you and your love as has the baby orang. Your desire to grant them all that they expect, and much, much more, is awakened to an irresistible degree.

Mrs. Benchley was not, however, as fond of chimpanzees—at least not at first. The first chimpanzees in the San Diego Zoo were a pair that came to Balboa Park through Dr. Wegeforth's father-in-law, who had a taste for exotic pets. Like many people before and after him, he kept them until they grew big enough to lose their cuteness and become self-willed. After the male, Bondo, had jumped his owner, Dr. Harry's father-in-law decided that they were dangerous and handed them over to Mrs. Benchley. In spite of trying, she could not learn to love them. The manners of the female, Dinah, who was coprophagous, offended her. Also, Dinah had the nervous habit of plucking hair from Bondo and herself to such an extent that both apes were sometimes nearly naked. Dinah ultimately died of a hairball in the intestines; later the management learned to feed its apes extra salt, upon which their destructive hair-eating tendencies subsided. Dinah's successor, Violet, was not very satisfactory either. In the end both surviving chimps were swapped for other animals, one of which was a young male chimp named Tim.

Tim was really young. When he arrived late one day to find himself all alone in his cage, he was unhappy; chimpanzees are very gregarious, and it amounts to real cruelty to keep them alone. Moreover, the man who took special care of primates happened to be away, and nobody else among the keepers knew how to persuade the new arrival into his sleeping quarters. Food did not tempt him to move because he had been fed all the way on the road to keep him quiet. Ignorantly one of the keepers tried to push him off his perch with a rod. This didn't work—Tim held on with all the strength of his hands and feet—so the man next tried to move him with a stream of water from a hose. Tim's resulting tantrum was violently noisy. Terrible screams came to Mrs. Benchley's ears, and she came running to find him high up on a perch, dripping wet and yelling his head off as he told the world all about his grievances. Furious, the director scolded all the men and sent them away. Then, going as close to the wire as she could get, she called the little chimp by name and talked softly and continuously. Tim looked at her, and then, sticking out his lips in a pout, he began to cry. As Mrs. Benchley went on coaxing him with her voice, he climbed down to her level and pressed against the wire. She put in her fingers and petted him; he seized them and held on for dear life. Finally she ordered the door to be opened and went in, whereupon Tim rushed into her arms, sobbing and telling her his troubles. Mrs. Benchley rubbed him dry with a gunnysack, gave him a bottle of warm milk and beaten egg, and stayed by his side until he fell asleep on his bed. All his life Tim never forgot that episode. In time he became king of his castle, with a mate named Katie and a little son. Whenever he went into a ranting display, as male chimpanzees are wont to do, and as our ladies were to learn, and Katie and little Georgie fled to safety, Mrs. Benchley, too, was wise and tactful enough to leave him alone and let him get on with it. But at all other times, even when he grew jealous, as he did, of the new gorillas, she realized that he had an eye on her. She was his standby, his special friend, forever.

Not so Katie. After Georgie was born, the director confessed in *Jungle*:

> To my great chagrin she refused from the first to permit me to come near him or touch him until he was more than one year old. She constantly warned him that I was dangerous and that he must avoid me, although she permitted Moore, her beloved caretaker, and Charley Smith, the head keeper, to play with him, examine his mouth and hold him in their arms in her presence. She apparently suspected me of ulterior motives and I was *verboten*. It was a situation I had never met before and I was both hurt and embarrassed by it. She would cover him, head and all, when he was tiny whenever I approached.

Fortunately George made up for this unfriendly behavior after he was weaned.

The fact that Mrs. Benchley took such an anthropomorphic approach to her anthropoids had made her a welcome guest at the Quinta Palatino, for Madam Abreu (the subject of the essay following) maintained a similar attitude. Mrs. Benchley had long been in correspondence with Madam before she visited Cuba, about the care and raising of orangutans, but on the day she and her fellow directors arrived, their hostess was preoccupied. She had just received two small chimpanzees and was busy letting them out of their crate and generally settling them down. Nevertheless, Mrs. Benchley was enthralled by what she could already see. At that time, she reported, the collection contained nearly fifty chimpanzees and orangs. "I will never get over the shock of sitting in her magnificent salon and watching through the dim light while the huge black attendant walked slowly up the beautiful marble stairs with a half-grown chimpanzee sitting on his shoulder and leading another by the hand." These particular animals, she was later informed by Madam, shared her bedroom every night, but all the apes were brought in from the dark every evening.

The two little newcomers had arrived in a large crate, in the charge of a man Mrs. Benchley described, disapprovingly, as loud-voiced and excitable, who kept saying that he could manage them perfectly. He had with him two little suits of clothes that he was trying to persuade them to put on. Now it so happened that if anything really annoyed and disgusted Belle Benchley it was animals dressed as people. In her view the little chimps were quite rightly reluctant to obey the man. They were tired, confused, and frightened, and when one of them spotted her, standing among the onlookers at the table, he rushed into her arms, closely followed by the other. It was then that Madam Abreu noticed her for the first time. "You are a good woman, the little fellows always know!" she cried.

The incident marked the beginning of a lifelong friendship and a vastly stepped-up correspondence. More than once Madam urged that the director return to Cuba or at least declare herself willing to inherit the chimps if anything should happen to herself. All this was due to the reaction of those two chimpanzee infants. Though chimpanzees were not her favorite species of ape, for some reason Belle Benchley got on well with them. She thought it might be some soothing quality in her voice that attracted them. Though a few chimps at San Diego, she flattered herself, were not so bad and reasonably well behaved, there is something in the chimpanzee makeup that she could not admire. Perhaps, one supposes, it is the incipient hysteria; chimps *are* excitable, easily pushed off the edge. How unlike they are to gorillas!

It was the gorilla species that really put the San Diego Zoo on the map. Very few of these great animals had been seen by the American public in 1931. Here and there, if one hurried, one might see a small gorilla on exhibition at the Ringling Brothers' circus, but they were not plentifully distributed through zoos. Only three lived in the States just then. Not many members of the public were even aware that a difference existed between the lowland and the mountain gorilla, but early that year word went out among the *cognoscenti* that the much publicized explorers Martin and Osa Johnson were the possessors of a pair of mountain gorillas that they would soon bring home from the Belgian Congo. These animals were young and had been in the Johnson camp for some months. Perhaps they were a true pair, perhaps not; it is very difficult to sex young gorillas. Competitive feeling was running high among the zoos. Before the Johnsons arrived, the news went ahead of them that the king of the Belgians, in giving his permission to the explorers to capture and carry away these rare animals, had got from them the assurance that they would choose the recipient zoo very carefully indeed.

It can be imagined how eagerly the San Diego contingent discussed the matter and how strongly Mrs. Benchley longed for the prize, but she didn't have much hope. The San Diego Zoo was among the youngest of the serious institutions, and San Diego itself was certainly not one of the United States' bigger cities. However, one can but try. Belle Benchley sat down and wrote a long letter to Martin Johnson, putting forward her zoo's claim as a worthy recipient of his animals. She described San Diego's system of family group displays, its roomy cages, its supply of fresh fruit and vegetables all the year round, its excellent record of health and longevity among the animals. She was delightfully surprised when, after a long, anxious silence, she got a telegram from Martin Johnson tentatively accepting her bid and naming his price for the gorillas, which was far less then she had expected: fifteen thousand dollars. This, although a large sum in 1931, was nothing compared with the asking price today, which is for lowland gorillas at that; mountain gorillas, being a threatened species, are beyond price. The zoo didn't have fifteen thousand dollars, but Dr. Harry, as usual, scouted around and raised it. As for the new cage that would have to be built for the gorillas, Martin Johnson donated that money. Mrs. Benchley must have written a powerful letter.

In the meantime, the San Diego construction manager was chaperoning the animals during their trip across the country, pausing now and then to send reassuring telegrams. And then, at last, they arrived: Ngagi the big one, Mbongo the smaller. They both proved in the course of time to be males, which was disappointing, but no matter; they were splendid creatures, dignified and beautiful. Mrs. Benchley fell completely in love with both of them. There is something about gorillas. She would not be the last to feel the fascination exerted by these massively dignified beings. Dr. Harry felt it, too, which was probably part of the reason he conceived the vision, shared by his loyal director, of someday creating a bigger place than the San Diego Zoo, a genuine wild animal park not too far away, where gorillas, among other animals, could live their lives out surrounded by trees and bushes and safe from hunters.

According to *The San Diego Zoo*, a pamphlet put out by the authorities, Mrs. Benchley brought bits of goodies to feed the gorillas every day—very likely, said one of the keepers, because she wanted them to be the biggest in the world. In vain the keepers expostulated and told her she was making them obese. She would not stop, and probably they did become obese, but there is no hard-and-fast word on it. What Mrs. Benchley wanted was to make friends with them, and she soon had her way with Mbongo. Ngagi was an older, cannier gorilla and held off for a long time. "You don't know what it's like to be snubbed," Belle said to a friend, "until you've been snubbed by a gorilla." In the end, of course, she had her way, and the gorillas became friendly.

In the early days of my acquaintance with the zoo world I attended a meeting of the American Association of Zoological Parks and Aquariums (AAZPA) in San Diego. On the last night of the conference we had a farewell dinner and were happily talking when there was a stir and a sudden stop in conversation. Everybody looked at the door, and my dinner partner said, "There's Belle Benchley!" There she was, a frail elderly lady with glasses and walking stick. There was a general rush to greet her. Everybody wanted to talk to her, everybody seemed to love her—and these were humans, mind you. I like to imagine what her reception would have been like if we had been apes from the zoo.

2. $\mathcal{M}$ADAM ABREU: COLLECTOR

LATE IN APRIL 1915 people who were not absorbed in World War I—and there could not have been many of those—noticed an unusual item in the newspapers—namely, that a baby chimpanzee had been born in captivity. This was the first time the civilized world had heard of such a thing happening, and scientists were the more interested to learn that the birth had taken place not, as might be expected, in some African rubber plantation but at Quinta Palatino just outside Havana, Cuba. Excited inquiry elicited the information that the apes belonged to one Madam Rosalía Abreu, a lady (it was said) of fabulous wealth and a strange enthusiasm for primates. She collected apes and monkeys, the bigger the better. This news, at a time when the public still thought of gorillas and chimps (which they usually confused as one and the same) as giant monsters, was enough to fascinate anybody. Who was Rosalía Abreu and how did she get that way?

The scientific reaction was calmer, but interest was just as keen. Dr. Élie Metchnikoff, the founder of primatology in Russia, was at the time at the Pasteur Institute in Paris, and he wrote to her immediately, asking many questions about the birth. Madam replied promptly and answered his questions in detail. He wrote again to thank her and included more and yet more questions: "Since you are continuing your observations perhaps it will be possible for you to tell me still more details of the life of your chimpanzees. What is the procedure by which the male arrives at intercourse with the female? Does she give him her love at once or does she first offer some resistance?" He wanted to know about the mother's treatment of the young and how the father behaved toward it and so on.

Madam replied to all the inquiries with unfailing good humor and, very likely, a natural pride. She told him the name of the infant's father (Jimmy) and described the temperaments of all the chimps. Then, no doubt, she turned to the equally exigent requests for information from Dr. Robert Mearns Yerkes of Yale. From her letters, which he later published in his aptly named book *Almost Human*, we learn that the baby was named Anumá, the Spanish version of Hanuman, the Hindu monkey god. She wrote: "The first time he reckanize me was the 5th of Mai, being only nine days . . . he look at me and scream hu! hu!" And she urged Yerkes to come to Havana and see for himself "my Cuban baby chimpanzee, which is the sweetest creature . . . that ever I have seen."

Though these two were destined to become colleagues of a sort and certainly friends, their actual meeting had to be postponed. What he was to discover about this Cuban lady was that her wealth made it unnecessary for her to be anything but herself—enthusiastic, religious in her own way, and very, very fond of monkeys and apes. Darwin's theories never troubled Rosalía Abreu. God made all of us, she thought, so why should we not be all close together? On her land at Quinta Palatino she had lots of room, plenty of workers, and all the food she could buy, so she set out to collect the animals she liked best in all the world. She liked all animals, but monkeys and apes were her favorites. It was no wonder, therefore, that her female chimp Cucusa and her fierce male Jimmy should have bred and produced the enchanting little Anumá. No other collection in zoos or private custody had the necessary animals to do such a thing, so it fell to Rosalía Abreu to make history.

Dr. Robert Mearns Yerkes was decidedly not rich enough to make such a collection, but he had something perhaps better than money: he had brilliance. The son of a farmer who had no use for higher education, he worked his way through college and on in postgraduate studies until he reached the heights of scientific academic circles, teaching first at Harvard and then at Yale. Philosophy, psychology, and biology were his subjects; he coined for himself the title "psychobiologist." He was nearly forty when he first wrote to Madam Abreu and was due to be enlisted in the American Army in the near future. For years he had found that objects for his special study of primate psychology were few in the States and very hard to find. At one time he had even moved with his family to California for some months to be able to observe an orangutan there. The thought of all those apes and monkeys in Cuba made his mouth water. Not that Madam Abreu seemed to take a scientific interest in her pets—her letters showed a

regrettable, if charming, sentimentality—but who else had so many primates? He longed to make a pilgrimage to Quinta Palatino, but he had to
wait until the end of the war. (During the war, as a matter of fact, he
invented the army's IQ test.) By the time he was out of the army and had
settled again to work, however, he himself had acquired two young chimpanzees, Chim and Panzee, and was thus on a more equal footing with the
chatelaine of Quinta Palatino.

In the meantime, what of Madam? What is her background, and how
did she happen to become owner of so many rare animals? In *Almost
Human* Yerkes would address the inevitable comment her unusual passion
attracted. He said:

> Madam Abreu's life has been such as to develop and strengthen the naturally
> strong will, which causes her on occasion to dominate any situation. Gifted
> with a fine sense of humour and a keen intellect, she thoroughly appreciates
> the behaviour of her unusual pets and the attitude of others towards her
> hobby. Only a person with rare independence of judgment, courage and
> freedom from conventional restraints, could have followed the course she
> has taken. . . . For, whether or not one likes the monkey kind, it is manifest
> that most persons are strongly prejudiced against those caricatures of human
> beings and would not accept them willingly as pets. . . .

She was born in 1862, eighth child in a family of seventeen. Her father
was a rich businessman named Pedro Nolasco Abreu. Rosalía adored him
and in her sketchy memoirs tells proudly how his mother had put him in
charge of affairs when his father died because she had so much confidence
in his talents—confidence that was borne out as the family fortunes
flourished. Quinta Palatino was their summer home. They spent most of
the summers there unless they went to the United States, usually to
Saratoga, New York, and the races. The family's all-year-round home,
however, was a plantation at Santa Clara, Cuba.

Rosalía could not remember when she was not crazy about animals, a
taste she shared with her father but not with her mother. As a child she
owned various pets—a cat, two birds—and when she was in her teens, a
little Mexican dog that her mother didn't like. One day, when Rosalía came
home, the dog was missing; her mother had taken it away and would not
tell where. For some reason Rosalía was sure that a certain cousin could tell
her what she wanted to know. He could but wouldn't tell her unless she

gave him a kiss; he was trying to marry her. So she kissed him and found out where the dog had been hidden. Now, how to get it back without infuriating her mother? Cunning was called for, but it wasn't really hard for a member of the rich Abreu family. There was a sister of Rosalía's married and living in France.

Rosalía told her mother that she had a strong desire to visit the sister, so that was arranged. Then she got in touch with another of her suitors, who was able to scrape up the necessary money and, with the dog, followed her to France. We are not told what happened to the suitor. Did she marry him? She married someone, but she never talked about him, at least to Yerkes. As for the dog, it lived on, at a safe distance from her mother, for twelve years in all. She owed her life to that dog, she told Yerkes, because once when she was very unhappy and had made up her mind to kill herself by jumping out of a window, the dog barked and licked away her tears, until, said Madam Abreu, "I came to my reason."

She was married in 1883, at the age of twenty-one. She and her husband stayed at Quinta Palatino for about a year and then moved to France, where all of her five children were born. As a schoolgirl, educated at convent schools in the United States, she had become fluent in English, and now she perfected her French. It was in France that she was separated from her husband, and finally divorced, about 1894. He is not mentioned by Madam in her account, and one does not know where he lived after she had left France.

That year, 1894, Madam was visiting the resort of Biarritz when she saw a little monkey, fell in love with it, and bought it, the first of her remarkable collection. This pioneer was, as she told Dr. Yerkes, a macaque, but what kind of macaque she does not say. She does say that she took it everywhere with her.

"When I came back," said Madam, "I went to New York, and stayed there to get my naturalization papers." Was Madam a U.S. citizen? No doubt whatever the situation was, it had something to do with the relations between Cuba and the United States. At any rate, after that matter was settled in 1899, she took the family back to Quinta Palatino.

In 1902 Madam bought her first chimpanzee, Chimpita, a half-grown male. The Abreus were at that time staying in Philadelphia (they had business interests in that city), and the chimp arrived just as they were starting for home in Cuba. So they took the animal, crate and all, on the train and then aboard the ship, into the cabin. Chimpita, Madam re-called, had cost five hundred dollars.

The ship was just under way when Chimpita got out of his crate and was free in the cabin. What were they to do? It was, remember, Madam's first experience with a chimpanzee. One thing she didn't want was a free-for-all with everybody in the ship chasing the ape. She thought quickly and grabbed an apple that fortunately happened to be within reach. She held it up for Chimpita to see and threw it into the crate. It was beautifully easy: Chimpita simply went in after it, and they slammed the door.

The second Abreu chimpanzee, the female named Cucusa, which cost three hundred dollars, came from Sierra Leone. Madam hoped that the two might breed; but they did not, and she became convinced that because Cucusa came from a country different from Chimpita's (we are never told where Chimpita did come from), he was not interested. Not, at least, in an effective way. They were of the same species. However, Madam may have meant that they were from different districts.

In 1908 or perhaps 1910, Madam was not sure which, she bought Jimmy in London from an animal trainer who had used him on the stage but could no longer manage him. Jimmy came all the way to Havana by steamer, alone and unchaperoned, to be met in Madam's absence by one of the Abreu sons. This chimp never became manageable, and unlike the other apes, he disliked Madam—she said the trainer had *told* him to hate her—but he was a good stud, so they put up with him. As for Cucusa, she and Madam got on very well. When she first arrived, she was covered with lice. Madam made a concoction of bichloride of mercury mixed with alcohol and rubbed the ape all over; it settled the vermin very quickly.

One day Chimpita and Cucusa, who were cage mates at the time, got out into the grounds and started running about. One of the guards, who, of course, was familiar with the animals, fired his revolver into the air to frighten them back into their cage. Chimpita reacted satisfactorily and went back, but Cucusa came up to the guard, put her hand in his, and took the revolver away. On another occasion the guards were moving the two chimpanzees from one cage to another, and Chimpita made good his escape. Madam told Yerkes:

I had a house full of guests. The guards were chasing Chimpita, I chased him, everybody chased him. One of the guards came up to him and tried to chain him, but Chimpita took the chain and threw it away, as if saying, "Liberty!" At last he climbed a high mango tree and wouldn't come down until I went close to the tree trunk and said, "Look, Chimpita, I have hurt my arm." He came down immediately to kiss me, and we took him without trouble.

She was convinced he understood everything said to him. One day she was feeding him, grape by grape. She never let her pets swallow grape seeds for fear they would get appendicitis, and as Chimpita ate, she held out her hand for the seeds. With his fingers and, sometimes, his lips, he picked them up and gave them to her, all except two, which had fallen into a little trench between the cage wire and the cement and which he couldn't reach. "Chimpita," said Madam, "you know perfectly well that when I am gone, you will eat those seeds." He gave a long-suffering look and went off to find a small twig. He brought it back and prodded the seeds out with it and duly handed them to her. She always claimed that Chimpita was her most intelligent chimpanzee.

Still, they were all, or nearly all, intelligent. Madam recalled how she had taken a female, Minina, into town one day for an X ray. Minina saw that Madam was wearing a hat, so she herself took the chauffeur's hat and insisted on wearing it until they had reached their destination. And another thing, said Madam: jealousy is a very strong emotion in animals. At the Quinta they had a male baboon that always hid his mate when men came to the cage to look at them. Because he never showed this reaction in the presence of women, Madam tried an experiment. She brought a priest to the cage side, thinking that perhaps the priest's cassock, like a long skirt, would fool the animal, but it didn't; he hid his wife just the same.

When Yerkes and his assistants finally arrived in Cuba, primed and ready for their summer's work, the Quinta Palatino collection consisted of almost eighty animals all told. Although chimpanzees were certainly the stars, there were many others, and not only primates; Madam's affections, as we have seen, extended to practically any member of the animal world. She mentioned in passing a family of bears, though she didn't care for them so much, and no wonder. "I heard the mother screaming," she reported, after the bear had given birth to a litter of at least three cubs. She went out to the cage and found that the male bear had eaten two cubs and had a third in his mouth. She went in with a stick and, with the help of her guards, separated the animals. "They are not such interesting animals as some others," reported Madam.

Then there was the elephant, a small Indian one she bought at long distance in Singapore. "Everyone says they are so bright, but I do not think so," she said. However, the children rode him everywhere, and she took him along on one of her trips to Europe, although this necessitated building a special addition to the ship. On the whole, she liked monkeys and chimps best because they are so intelligent and affectionate. She said that she didn't find orangutans as interesting as chimpanzees, perhaps (one suspects) because at Quinta Palatino they were not very active. Two orangs from Borneo died there. But about a month before the Yerkes contingent arrived an orang escaped and had to be chased from tree to tree, swinging along. "He is a beautiful animal," Madam admitted. They brandished sticks at him until he was frightened enough to descend. "He was very tired and scared," said Madam, "and he wouldn't look at anyone because he knew he had done wrong."

When Yerkes asked Madam to tell him in detail about the famous birth of the infant chimpanzee Anumá, she decided to let the guard give his version as he was present at the whole affair. He did so, and she translated: "In the morning she had labour pains, and rattled the chains to have them taken away." This is rather mysterious. Evidently all the big chimps were chained to some extent, at least sometimes.

> When she was unfastened from the chains she lay down and Andrés [the guard] went away. When he returned after a few minutes, the little animal had been born. We thought it was dead. Then the mother commenced to breathe into his mouth and drew the tongue out from the mouth with her lips. Then she commenced to clean the infant, and when it was dry she began to chew the umbilical cord: and when she came to within a few inches of the infant she closed the end by chewing. She took the placenta, broke it, and ate the blood, leaving the skin behind.

How did Cucusa know what to do? Had she seen the birth of a little chimpanzee in the wild, before she was captured? Nobody knew the circumstances of her own infancy.

Yerkes asked Madam many questions, and she seemed more than willing to impart the knowledge she had collected throughout the years. The period of gestation for chimps, she said, is nine months. She almost had it right; it is nearer to eight and a half. She measured the time, however, from the last menstrual period; that probably accounts for the difference. She had a strange story about one of her females, Monona, who "knew" that she was too young to bear a child. Every time a male mated with her, Monona would go and wash herself, until Madam, eager to breed her, gave orders to keep water away from her. "So she had her baby, but it was too small," Madam admitted. "Females should be ten years old before breeding." They are probably ready, however, at eight or nine; males are mature at ten. Females have grown their permanent teeth at eight or nine; the milk teeth begin at four months. When their milk teeth dropped out, Madam's chimps handed them to her as if to say, "These belong to you." The molars are the last to grow.

Madam declared that animals must have fresh air and trees around them or they will not breed. Parrots are the same, she asserted; she had parrots that did not breed until she took them to France to a warm climate. She didn't say where or what part of France could be warmer than Havana. In any case, the birds laid eggs there and had done so since they came back; they had hatched young in Cuba. If conditions are not right, said Madam, animals won't reproduce because they know their children would not be happy.

Once Yerkes asked if she thought monkeys of different countries—i.e., species—would not marry, as she put it. Oh, yes, sometimes they do, said Madam. They had one case at Quinta, and both the mother and infant died. Generally they don't mate between species, but exceptions prove the rule. Her lion-tailed monkeys, however, and her green monkeys never paid attention to any but their own females. Yerkes reminded her that humans, blacks and whites, do crossbreed. "Not so much anymore," said Madam. "Not in Cuba."

Madam believed firmly in the soul and seems to have extended this belief to the nonhuman animal kingdom. Cucusa, Anumá's mother, died during a visit she made to Europe with Madam. Both she and Jimmy were there when it happened. Cucusa knew she was dying, Madam said. She put up her hands and cupped Madam's face, kissed her good-bye, and expired, while Jimmy, in his cage outside the cabin, screamed and screamed. "He looked past me, as if he could see something I could not, and his lower lip hung down. It made my flesh crawl," said Madam. "He did the same thing when others died. Five times he has done it, the same thing." Yerkes asked if she thought a soul, after long sojourn in a chimpanzee, might enter a man. No, Madam didn't think a monkey could become human.

This conversation probably held special interest for Robert Yerkes because of a tragedy that had befallen his entourage there at Quinta Palatino. Prince Chim, his brilliant little pet, suddenly sickened with pneumonia and after two weeks died. The little ape's behavior during the sad last days might well have been cited by Madam in support of her ideas. "Often as I entered the sickroom," wrote Yerkes in *Almost Human*, "I noticed that he was watching eagerly and although he welcomed me and sometimes even came to me from his bed, he seemed still to be expecting some one. Every footstep in the hall attracted his attention and it was almost as though he anticipated the appearance of some one or some thing not present. . . ."

Leaving the subject of anthropoids for a while, Yerkes asked about marmosets, those very small New World monkeys that almost always give birth to twins. "Ah," said Madam, "I must tell you about them. An animal dealer once told me that when the female dies, the male dies soon after, and when the male dies the female dies. It is very true. I have lost about eight, some in family ways, some because they could not have a baby. Some died after the baby was born. Every time the male died, three or four hours immediately afterward, the female died. I had twenty: I have six now."

There is little information about marmosets, Yerkes commented. "What are your impressions? Are they intelligent?"

"In their way," said Madam. "They are very affectionate: I have seen some that were *very* affectionate. But you must have them on a little cord, always with you."

Why should so many die while giving birth? Because, said Madam, their pelvis is so narrow. He asked: Might it not be because of crossbreeding, so that the infants are too large? Answer: No, they marry only within their own species. What, in her opinion, is the monkey closest to the chimpanzee in intelligence? Answer: the Panama squirrel monkey, the one with the black cap of hair, *not* the Brazilian squirrel monkey, which is very stupid. And so it went on, day after day.

Yerkes and his assistants kept careful logs on the animals they watched. On one occasion Madam wanted to try out the effect of sudden loud reports, and she had her assistants watching as the chimpanzees dashed about, some frightened and some not, depending on their natures. The Americans tried going into the cages of some apes, although never Jimmy's. They watched, and made notes on, other primates, such as the baboons, of which there happened to be one couple at the Quinta. But there had in their time been more baboons—as many as ten. When Yerkes asked Madam why she had kept so many if she didn't care for them, she replied, "Because they were born here and they are my children, poor things."

He asked her if she had any anecdotes about baboons, and she said:

Yes, I have one very tender story. The original male, who died not long ago, lived to be about twenty years old. Towards the end he was very sick and helpless, and one of his babies, not quite a year old, took care of him, giving him food and helping him around, when none of the others paid any attention to him. The wife and mother neglected him, but the baby would not leave him. That is a very tender story. And I will tell you something else, that same baby got out of the cage one day and escaped. We found him in the overseer's house, and had to put him back in the cage through the same hole he had used to escape through. Then the guard put a piece of wood over the hole. The minute the baby baboon was back, his parents looked at him very carefully—up his nose, into his eyes, in his mouth, everywhere, and when they found he was unhurt they seemed greatly relieved, and came to us to say, very nicely, thank you. They are stupid, but they love their children.

Yerkes wrote:

At Quinta Palatino there were in July, 1924, eighteen anthropoid apes, including one Wau-Wau gibbon, *Hylobates leuciscus*; three orangutans, *Pongo pygmaeus*; and fourteen chimpanzees belonging probably to two or three different species. These creatures constitute the very heart of the collection, because they loom large in bulk and are ever present to one's vision by reason of varied and interest-compelling activities. All of the primates are eager for human attention when held in captivity as tamed creatures, and it is inevitable that the larger and more demonstrative ones should have great advantage in competition with their smaller and less gifted relatives.

They are so highly individualized, continued the writer, and so quickly make a place for themselves in one's world of social relations that one can't describe them by merely naming their type—gibbons, chimps, or whatever. Each of Madam Abreu's apes had a name to which it responded "and which rapidly gained significance for the frequent visitor." To prove it, he proceeded to name them and described them, starting with Wau-Wau, the silver gibbon, a beautiful creature with hair so thick and soft it was like a fur covering. Wau-Wau was alone of her race at Quinta Palatino and so was lonely and tended to be timid and wild. Madam didn't think she was very intelligent; gibbons, she said, are rather like the langurs of India. Certainly they look like langurs, except for being tailless, but it is difficult to agree with her as to gibbon intelligence. However, if her standard of comparison is to measure animals up against human mentality, she is probably correct. A gibbon is not intelligent in the way a chimpanzee is, for example. A gibbon, also, is not imitative, as a chimp is. A man I knew in Shanghai described my Mr. Mills as ingenuous, and that is a good word. When he meant to grab or steal something, his round eyes fixed themselves on the desired object for a good long time before he grabbed it. Even more interesting, you could see him summing up his approach and escape, for that is the way a gibbon plans his journey through the trees, carefully, even scientifically. Now a chimp would not do that. You can't tell from the way his eyes work what on earth he is planning.

It is customary, noted Yerkes, to judge an ape's (or a monkey's) so-called intelligence by the aptitude it shows in manipulating objects—in other words, tool using. Chimps have long been known for adapting rods, barrels, or anything else they can lay their hands on to their own use— fishing for things to eat, piling barrels on crates to increase their height, and so on. Monkeys have not shown themselves so adept, but here and there one does play with a toy. (If Yerkes's observers had been patient, they might have discovered that baboons, at any rate, are good at catching balls.) There were no chances to observe gorillas; Madam's efforts to obtain one of those rare animals had never been successful. But there were three orangs at Quinta Palatino, and in comparison with the chimpanzees they seemed very backward.

On another point, however, Yerkes found himself sadly at odds with Mrs. Benchley and her opinion of gibbons. "The voice is disagreeably penetrating," he wrote, "and even in the open is an unwelcome intrusion if one is near by or if several of the animals cry in chorus."

All these observations and tests were exceedingly valuable to the anthropologists-cum-biologists-cum-psychologists who made up the Yerkes team; but their leader thought that his main study was Madam herself, that interesting primate collector, and he did not cease to observe her and hold long conversations with her when circumstances permitted. He had his own ideas, of course, as to how the collection might be better managed. He asked her once, at least, why she didn't give the animals more things to play with.

"Well, what?" she asked. "I can't think of anything but trapezes and tires and they have those. They break everything else." All zookeepers know the feeling.

But Yerkes did not complain that the mistress of Quinta Palatino neglected her charges in any other way. He described the surroundings and the regime:

> A great point is made at Quinta Palatino of the proper combination of sunlight and shade. Cuba has an abundance of both, but sometimes they are difficult to regulate. It has required foresight and careful planning properly to design, locate and construct the ape cages. They must be placed in proper relation to the trees which in the several seasons are expected to provide natural shade, but at the same time they should be so located that the chill winds of winter, or of late afternoon and night, of cloudy weather and of sudden storms, shall be tempered by natural or artificial barriers. There must be free circulation of air throughout the cage and the sleeping-room, and for a portion at least of each day the animals must be able to retreat into the shade, for to be compelled to lie or sit in the direct sunlight is undesirable and may be dangerous. The cages therefore are provided with roofs which turn both water and sun.

The animals had to have a spacious and secure enclosure of their own, because they were so near Madam Abreu's house, and a ventilated room in which to retreat from the cold and spend the night. They could be trained to sleep in hammocks or on mattresses and to cover themselves—as, indeed, many do in the wild with leaves or branches. But in one respect Yerkes found the cages deficient. He thought there should be a sort of communal dining room, to which all the animals should be admitted for meals and where they should be trained to behave like well-trained children, "eating what is provided and returning to the outdoor cage when the meal is finished." Of course, he added blithely, such a program would require many careful, competent attendants, and intensive training and discipline of the animals themselves. One wonders what Madam's feelings were when her guest made this suggestion, which, if put into practice, would have called for at least one "attendant" for each animal. To that date, Yerkes continued, Madam Abreu had not taken this step in the domestication of her pets. Instead, each individual or group was fed in its living cage, and he considered it a pity. He wrote:

In doing so, she misses an important opportunity to vary the routine of the animals' existence and to afford them a chance for new adaptations and new adventures. At the same time, the humanizing of eating would have the very real advantage of assuring to each individual, be it weak or strong, its own proper portions of food. Cage feeding means that each individual scrambles for what it desires and that some get more, even if none gets less, than they need.

If he were to do it, he said—well, he did say "we"—he, or they, would suggest a special dining room or dining cage near the other cages, with a long table and chairs and with facilities, also, for use as a playroom or schoolroom as well.

The cages were so placed that the animals could readily see what was going on in the others, an arrangement peculiarly important to the great apes. (The Bronx Zoo followed the same plan fairly recently while remodeling the Great Ape House, especially in the room where gorillas are caged. The main idea is to help them understand breeding and caring procedure.) Another acutely important part of keeping animals is their diet. Madam's routine in this respect was therefore interesting, because she had a lot of experience in successful feeding.

"The routine varies somewhat as to hours, with the season of the year, but the following is applicable to spring, summer and early fall," he wrote. At 7:00 A.M. the animals got bread or cooked cereal and milk. At 9:00 A.M., if mangoes or other native fruits were available, they all had some. About 11:00 Madam herself made a round of the cages, giving to each individual or cage group bits of cooked food from her table or any extra delicacy available. She was able to greet each animal in friendly fashion and pet it—everyone, one assumes, but Jimmy. At 3:00 P.M. the animals got the last regular feeding by the keepers: usually cooked food, including white or sweet potatoes, squash, and corn, sometimes green on the cob, as well as baked plantain—whatever was in season. They also got milk. The food they got was rather like that eaten by humans, with the exception of meat. Nonhuman primates have likes and dislikes, preferences and prejudices in their food, just like us, and if it's possible, it is as well to observe them. Occasionally an ape would eat egg (cooked or raw) or bits of meat. Some were reputed to eat lots of meat, even to live on it, but Madam didn't think they needed it. She didn't give any meat at all to the gibbon, the orangs, or the chimpanzees, and they did well without it. After all, there are human vegetarians as well, and they seem to thrive. As for what the apes liked, Jimmy adored pineapple; his mate Monona preferred bananas, as did Fifille and Lu Lu, although like Jackito and Blanquita (these all were chimps), they were also fond of oranges and mammee. The monkeys were fond of most fruits but seemed to prefer oranges. All the animals in the collection liked corn, either green or cooked. The baboons preferred hard corn, palm nuts, and sunflower seeds. All loved coconuts, both the milk and the pulp. Madam Abreu didn't give the chimpanzees or the orangutans hard grain, for fear of stomach trouble from improper clearing. Chicken fed to the gibbon caused serious bowel trouble. Pregnant females got extra rations of milk, as did the babies. Pure water was supplied three times a day, as water in the cages tended to become dirtied.

The attendants were careful not to overfeed and not to leave food lying about. Today any good zoo director would agree with these strictures, but Madam Abreu found out the hard way, by experience. Yerkes, an idealist in certain matters, felt that they could be improved.

"Undoubtedly the ideal method," he said, "is to feed the animals individually, supplying to each just what it can eat readily and to advantage, precisely as in the case of a person. The animals can be trained to eat neatly from food containers and even to use a spoon in feeding themselves. The food can be offered then by courses or in such other way as is appropriate and the whole procedure conducted in an entirely orderly manner."

Madam always took some of the animals into her house for the night, the orangutans and some of the chimpanzees included. There they were securely caged. She did this to keep them from taking cold or contracting pneumonia during the cold nights. To every animal in the house at night she gave milk at eight o'clock or, if she was out for the evening, when she got home. In the morning between six and seven they were returned to the outdoor cages. Only Jimmy among the chimps was never brought into the house but was fastened in his bedroom for the night and let out early in the morning to his large cage. Anumá and Monona, though large, were not excepted from the rule; every night they were taken to their cages in the house. Irrationally enough as it seemed, the gibbon stayed out of doors. Like the monkeys, she went to sleep at sundown and awoke at sunrise and remained a very healthy gibbon. Yerkes doubted if all the business of taking the animals indoors was quite necessary. He pointed to Jimmy, who was perfectly healthy in spite of staying out all night. "We are inclined to believe that their owner humors the animals somewhat unnecessarily," he said, "at the same time indulging her own sympathetic interest in them and her eagerness to be good to them."

From the psychologist's point of view, one of the main problems to be faced in dealing with captive animals is that they are in some cases cut off from their own kind. At Quinta Palatino most of the animals had company—although the gibbon had not—so the hardship was somewhat mitigated. It seems clear, however, that animals, especially primates, can make of their human caretakers companions that take the place of members of their own species. "Presently a monkey or an ape may come to behave towards a person much as it would toward one of its kind, acting to attract attention, seeking in various ways to share in activities, above all, using various tricks to encourage play relations," he wrote.

"Animals are more bright than people: be sure of that," said Madam. They were talking again about souls, and Yerkes asked what might happen if, for example, he had a bad soul and it was still bad when he died. Madam said:

> For the sake of justice, I put myself in the place of God. I say, "I will make this thing with my own hands." Well, I make it, with sentiment, blood, and nerves; with sensation, with brains, with life, with every kind of pain. Will I make this animal and then, when it dies, this finishes everything? No, I will not do it. Then will God, who made me, do it? That is the reason I feel [animals] have a soul. They will be recompensed, surely. As for heaven and hell, I cannot believe that God will make for you, the worst man in the world, forever punishment for a thing that it does temporarily. They say, eternal fire. Fire for a soul that has no body—it is ridiculous. I say, see all those stars there? They cannot be there for nothing. Very well, why will not God make people come from one planet to another? Perhaps the soul goes into the body of a little child not yet born. Until my soul is perfect I will not go to God—and where is God? Perhaps you will laugh. I say God is in the sun, because it is life. The soul will migrate from one body to another—an animal, perhaps. Perhaps God will send you into an animal, and you will have to suffer.

Yerkes asked, "Do you think that will happen in the same world?"

"Perhaps in another world," said Madam Abreu. "Of one thing I am sure, there is God, sure as I am here. There is God and there is Providence.

Sometimes I have doubts, but then I say, what am I? A little piece of earth? I cannot see the plan."

"What do you suppose becomes of the animal's soul after it is perfected?" asked Yerkes.

"Ah," said Madam, "that is another thing. I always ask Him to be good to the animals. Then I say, 'Why do I ask you for something for those that you have made? We have been here together, and may we not go on together?' "

Madam Abreu died in 1930.

3. $\mathcal{A}$UGUSTA MARIA DAURER DeWUST HOYT AND TOTO

Toto and I: A Gorilla in the Family, by A. Maria Hoyt, tells of a truly personal involvement with a gorilla. The book is an account, published in 1941, of how a female lowland gorilla came into the possession of Mrs. Hoyt and her husband, Edward Kenneth Hoyt, and changed their lives. The Hoyts married in 1926. From the fact that Mrs. Hoyt's name at the time of that marriage was Mme. Augusta Maria Daurer DeWust, I presume that she had been married before, but she says nothing of this. In fact, about all that she says of her past is that she was born in Argentina. Otherwise, she relates these pertinent facts: that her husband retired from business at the time of their marriage and, until Toto changed everything, devoted himself to his favorite pursuit, big-game hunting trips, on which Mrs. Hoyt always accompanied him. Yale alumni records show that in 1930, for example, the couple went on a four-month expedition in Kenya, Uganda, the Congo, and the Sudan and that their bag included elephants, two hippo bulls shot from a canoe on the Nile, and other game. Hoyt gave a number of specimens to the American Museum of Natural History and to the National Museum of Natural History, Paris.

In 1931 they went to French Equatorial Africa for more shooting, specifically of gorillas, of which Hoyt wanted a really big specimen for the American Museum of Natural History. The morning the involvement with Toto began, Hoyt went out with guides to pot his big ape from a group that African guides had told him was some distance away by foot and canoe. Some of the Africans had already made a clearing all the way around the group of trees where the gorillas were sheltering, and everything was apparently ready. The "boys" described the group as a family, with one

41

large male in charge and a number of females and young. Mrs. Hoyt went along, simply because she always did, but she stayed a short distance from the field of battle. Kenneth Hoyt shot the big male and with Mrs. Hoyt started measuring it, assuming that the rest of the group had escaped. He was wrong. The Africans, who loved to eat gorilla meat, had left vine nets all around the copse, and once the gorillas got tangled up they speared them all, with the exception of one female, too small to make good eating. As the Hoyts stood looking in horror at the carnage, one of the guides, a village chief, brought the baby gorilla, struggling and biting, to Mrs. Hoyt, who put out her arms to it. At that moment the little creature's behavior abruptly changed; she cuddled down in the woman's arms and clutched her. That was that. Like the dead gorillas, the Hoyts were hopelessly entangled in a mesh. Admittedly at the time Mtoto (Swahili for "child") seemed to become just another pet among various pets that the Hoyts had accumulated—a young leopard, a chimpanzee, an African eagle, five or six monkeys, and a pair of vulturine guinea fowl—but from the start she was not treated quite like the others. They lived in a "menagerie" compound at camp, whereas Toto slept in Mrs. Hoyt's tent. Her adopted mother hastily got busy cutting up her husband's flannel shirts and making them into bellybands like those worn by Argentine infants. Toto wore bellybands for months, and whether because of this or in spite of it, she survived the rest of the trip, though two gorilla babies subsequently acquired by the Hoyts on the safari did not.

Back in Europe, the Hoyts and Toto roosted briefly in Paris in a hotel on the Rue de Rivoli with Mrs. Hoyt's mother, her two Pomeranians, Mrs. Hoyt's personal maid, and Toto's personal attendant, Abdullah, who had come with her from Africa. Toto's every wish was granted, Mrs. Hoyt recorded, and she grew spoiled. "Have you ever seen a sickly child whose parents spent all the years of his early life fighting for his existence, who showed the signs of discipline which mark the training of a normal child?" Mrs. Hoyt wrote. This referred to many bouts with dysentery and one really serious attack of pneumonia suffered by Toto, during which the best pediatricians in Paris had interesting experiences treating her. She was under an oxygen tent for a while. Irritation, anger, and the like shown by the people around her "tended to upset the delicate balance" of the "baby's" nerves, Mrs. Hoyt wrote, so emotional outbursts were avoided as much as possible. After the pneumonia Mrs. Hoyt took the "baby" to the French seashore for convalescence, and the book contains a photograph of the two on the beach—Mrs. Hoyt very fetching in a big hat and a flowing white dress and Toto very small, black, and appealing in her arms. Toto may have been wearing jewelry at the time, though one cannot see it. She was fond of bracelets and necklaces; a later picture, taken when she was much bigger and very much fatter, shows her with a favorite bracelet on her hairy arm. While his wife and baby gorilla frollicked on the French sands, Kenneth Hoyt was in the United States, looking for a place where they could live, in a climate suitable for gorillas. In the end the couple decided upon Cuba, where the climate resembled that of French Equatorial Africa and where there was a kind of tradition about apes. Madam Rosalía Abreu, as we have seen, had maintained a colony of monkeys and apes near Havana. Madam Abreu had died in 1930, and her son had dissolved the primate colony then; but Havana remembered it well.

The Hoyts rented a spacious house in Havana, with gardens and a swimming pool, and moved in. Toto was ten months old and weighed twenty pounds. Soon she was beginning to walk upright and cutting her baby teeth. Mrs. Hoyt made a fuss about the first tooth, and thereafter Toto was often to be seen studying it in the mirror. One day she fell and broke off a bit of the tooth—a mishap that upset her very much. She ran to Mrs. Hoyt, mouth open, finger pointing to the damage. She grew and grew. She outgrew her first playroom, and the Hoyts built another, equipped with ladders, swings, and bars. After eight months the Hoyts decided that Havana did agree with them (and with Toto, too), and they bought the place they had been renting. Toto learned to draw with chalk on the flagstone walk, producing strange hieroglyphics that Mrs. Hoyt professed to find meaning in. Then came a problem: Abdullah had to go home. The Hoyts, like most affluent parents, had never coped with the real job of taking care of the baby. It was Abdullah who had fed and washed her and slept with her each night. Five keepers came and went during the three weeks that followed Abdullah's departure, and life at the villa was hell. Then Mrs. Hoyt heard that the late Madam Abreu's head keeper, a Span-ish-born man named José Tomás, was at liberty—after all, jobs looking after apes were few in Cuba—and she sent for him. Toto tried once, but only once, to scratch him. Tomás, Mrs. Hoyt wrote, held her "little hands" and scolded her until she was abashed. After that he almost always had her under control, and he remained with her for the rest of her life. There are many photographs of Tomás and his ward, in which the gorilla appears larger and larger. Inevitably in one of these Toto is standing on Tomás's back. I say "inevitably" because, as I have found out, gorillas, for some mysterious reason, like to stand on people. For the moment let me say that to me the most remarkable thing about the photographs is that Tomás, though he is usually in shirtsleeves, is almost never without a tie. Few men would dare wear a tie when they are close to a full-grown gorilla; the likelihood of a casual garroting is difficult to ignore.

Toto grew so rambunctious, even with Tomás in attendance, that the Hoyts at last built her a new, reinforced house, which was, to be honest, very much like a cage. She did not spend all her time there, but the place became more and more necessary in times of stress. To judge from the pictorial record (though no one would ever have dared say so to Mrs. Hoyt), the baby was growing obese, and no wonder when we consider her daily intake:

7:30 A.M. Culture bacillus in sugar water

8:15 A.M. One and a half quarts of milk with tapioca

10:00 A.M. Large mug of orange juice, carefully strained (she wouldn't drink it otherwise)

11:30 A.M. Eight bananas with cream cheese

1:00 P.M. Lunch, with either steak or chicken

4:00 P.M. One and a half quarts of milk with oatmeal or Pablum or Cocomalt

6:00 P.M. Two baked apples and a mug of milk.

So the baby continued to grow. Whenever she escaped the surveillance of Tomás now, he would blow a whistle to warn the household, and everybody would run for cover. Toto's first act in these situations was usually to rush for the main house and try to get in through a door or a window. If she got in, she particularly liked to mess up the laundry, tearing up sheets and soiling things. Mrs. Hoyt learned from experience the point at which to stop playing with her; when Toto got really excited, she was apt to pull Mrs. Hoyt's clothes off. Spare costumes for Mrs. Hoyt were always kept in a closet in Toto's playroom, just in case.

At the age of four or five Toto made a pet of one of the household kittens and carried it with her everywhere. A photograph shows her with the cat sitting on her head. In another, she sits on a garden chair too small for her, clutching the cat. In this picture Toto's body is enormous in both directions, and her face, which seems unusually round, is surrounded by a fine set of muttonchop whiskers. The baby was obviously growing up. In fact, even Mrs. Hoyt could see that Toto's character was changing. Her behavior pattern followed a definite rhythm. Every month, Mrs. Hoyt wrote, for two or three days during the new moon, she was unruly, hiding in the hedges and refusing to be caught, even by Tomás. The Hoyts devised an electric prod that would give her a shock, and after one experience with it Toto obeyed at the mere sight of it. But the need to use it was, as Mrs. Hoyt noted, a warning of "stormy times ahead." (Tomás also took to carrying around with him a live snake; Toto, like most gorillas, was terrified of snakes.) Tomás was just in time one afternoon to prevent Toto from braining one of the Pomeranians with a large rock. Toto did not care for the Pomeranians.

It is not hard to fill in the gaps in Mrs. Hoyt's narrative. She tells us that her husband, after Toto had thrown him down several times within a short period by yanking his necktie, brought up again the "sad necessity" they would someday have to face, of finding somewhere else for Toto to live. Mrs. Hoyt refused to think about it, even after Toto had sportively grabbed one of the Hoyts' employees one day and carried him toward the top of her cage, obviously intending to drop him from the highest possible point. Fortunately she more or less lost interest and dropped him from a mere five feet. He did not bring suit against the Hoyts, but he might have; worse, he might have been killed. Mrs. Hoyt insisted that Toto merely loved to make people run, "like a healthy puppy, charging into a flock of chickens just for the fun of hearing them squawk and seeing them scamper." Her husband was not convinced. In 1937 Mrs. Hoyt was still going out to Toto's house every afternoon to play with her. One afternoon, in the middle of a frolic, Toto jumped onto her swing and pushed herself with all her strength at "Mother," who, being knocked backward onto the flagstones, broke both wrists. She never dared tell her husband exactly how it had happened. She concocted some story; but he was not deceived, and worry over Toto continued.

In the summer of 1938 Kenneth Hoyt died in a New York hospital, leaving his grief-stricken wife to return alone to Havana and the "baby." Toto made her feel worse, she wrote, by looking everywhere for Kenneth. Yet Toto was a comfort, too; they would walk and walk together in the gardens, with Mrs. Hoyt talking to the gorilla, confiding in her. Photographs in the book show dramatically how heavy, literally, Mrs. Hoyt's responsibility had become. In one we see Toto sitting at a lace-covered table in the garden sucking Coca-Cola through a straw—a hulking figure, thickly furred, with a muscular chest, a fat belly, and a thick, short neck. In another Mrs. Hoyt, standing, pours tea, with a smile, while Toto, balancing on a little ladder, holds out her great, hairy hand for a cup. Pretty Mrs. Hoyt looks fragile and tiny and thoroughly incongruous near the gorilla, whose head in profile looks about three times the size of hers. On another page we see the two strolling in the garden, Toto's huge body on all fours. Her enormous face looks grave, and so human, though outsize, that the whole thing appears wrong—as if some person were imitating a gorilla and not doing it very well. Mrs. Hoyt smiles down maternally at her companion; this is the most natural thing in the world, she seems to insist.

By this time all of Mrs. Hoyt's friends were urging her to get rid of Toto, and before long a representative of the Cuban government came to her with a similar message and a threat: If she didn't get rid of Toto soon, the animal would be confiscated. To all this, Mrs. Hoyt turned a deaf ear, though the new-moon periods were getting so difficult that she scarcely dared leave the house. On one occasion Toto was angered because the household station wagon was blocking her way. She picked up the rear end of it and—though the brakes were on—pushed the car violently into the rear wall of the garage, shattering the headlights. Even between new moons, if her meals were as much as two minutes late, she would get away from Tomás and rush up to the servants' dining terrace, to throw plates, cutlery, and food against the door of the kitchen, into which the help had fled. Occasionally she threw the table, too. "It had become practically impossible for me to have guests at tea," Mrs. Hoyt reported, and the grocery boys would no longer deliver to the door; they threw the grocery packages over the wall and ran for their lives. The breaking point was reached one night when Mrs. Hoyt, who had gone out for the afternoon, stayed away and went to the movies with a friend. She came home in pelting rain, after dark, to find Toto careering all over the garden, ignoring Tomás and thoroughly drenched. To make Toto pay attention to her, Mrs.

Hoyt had to get one of the servants to pretend he was attacking her in Toto's room. Toto came at them like a locomotive, and the servant got out of the way just in time, slamming the door as he escaped, and all was well—for the moment, though Toto shook the bars in rage. This time Mrs. Hoyt had to confess that she was beaten: Toto had to go or be confined permanently. And this brings us to the story of another gorilla brought up by a woman, a gorilla with which Toto's life was soon to be joined, at least in a manner of speaking.

The other gorilla, a male, came to be known to the world as Gargantua the Great, but in infancy his name was Buddha. Gertrude Davies Lintz, the English-born wife of William Lintz, a Brooklyn physician, was a typical Englishwoman in at least one way: She loved animals. She bred Saint Bernards and showed them professionally, but she liked primates, too—especially the anthropoids. A sea captain friend of the Lintzes' named Arthur Phillips made a regular run to Africa several times a year in his ship, the *West Key Bar*, and he usually brought back a few exotic animals to sell. In 1928 he sold three chimpanzees to Mrs. Lintz, and three years later, when he came back with a sickly little lowland gorilla, she bought that as well. Soon she had nursed it to health and named it Massa. (I will discuss Massa at greater length in the following essay, on Gertrude Lintz.) Massa was wrongly thought to be a female. Much later Mrs. Lintz sold him to the Philadelphia Zoo, where he lived until he died at fifty-five, a record age so far for members of his species in captivity. (In his infancy, thanks to Mrs. Lintz, he survived both pneumonia and a disease that resembled polio.) Late in 1931 Captain Phillips put more chimps and another little gorilla aboard his vessel for Mrs. Lintz. But when he returned to the United States, he had a sad story to tell about the new gorilla. Until the ship sailed into Boston Harbor, the gorilla was in the best of condition. At that point a sailor who had just been sacked by the captain revenged himself by squirting a fire extinguisher full of nitric acid at the animal, grievously hurting it. The gorilla's skin was badly burned, and so were the muscles around its face.

Mrs. Lintz took the situation as a challenge, and, with the help of an assistant named Richard ("Dick") Kroener, went to work to save the gorilla, which, among other things, couldn't close its eyes. Buddha, called Buddy, was characteristically stoical about the treatment she gave him, and his eyelids soon healed. He was to carry other signs of this ordeal for the rest of his life, though. The left side of his mouth was fixed in a fierce grimace that exposed the teeth and gave him an appearance of ferocity. This was completely deceptive; he was an unusually sweet-tempered creature. He played, ate, and slept with Massa in separate parts of a room in the Lintz house in Brooklyn, gaining 100 pounds during his first year there. Mrs. Lintz has written in a memoir that she rated gorillas superior to chimpanzees. She liked their dignity, reserve, and independence. She may have changed her mind—temporarily, at least—after bringing her animals home from the Chicago world's fair, in 1934. Massa had learned to scrub the Lintzes' kitchen floor, and while he was doing this one morning, Mrs. Lintz, entering, slipped and fell, knocking over on him a pail filled with water. The gorilla, which, at five, weighed 140 pounds, was startled and bit her severely; a woman friend who was visiting the Lintzes ran into the kitchen and beat him with an iron skillet to make him let go. Mrs. Lintz was laid up for months, but even so, she held on to Massa, hoping the gorilla would quiet down. Finally, at the end of December 1935, she loaded Massa into her car and took him to Philadelphia, where the zoo, still believing him to be a female, had bought him as a mate for its male lowland gorilla, Bamboo. The error was discovered some months later.

Buddy, however, remained tractable, and Mrs. Lintz carted him around to shows for two more years. Then she began to wonder whether he might not turn on her one day in the same way Massa had. The attitude of her helpers may have influenced her; as Buddy grew bigger and stronger, they lost their insouciance in dealing with him, and these things are catching. One night a bad thunderstorm swept over the city. Buddy was at home in Brooklyn in his cage in the Lintz basement, and the Lintzes were asleep upstairs. Though Buddy now weighed four hundred pounds, he was still afraid of things, and the thunder scared him. He wanted his "mother." He got out of the cage (someone had forgot to lock it fully), tiptoed upstairs, and crawled onto the foot of the bed Mrs. Lintz was in, pinning her down. She got a horrid start when she became aware of Buddy huddled there, sobbing. But she kept her head. Talking to him softly and soothingly, she led him by the hand down the stairs, through the basement hall—where she grabbed a pear for him from the sideboard—and on to his cage. She tossed the pear into the cage, Buddy rushed in after it, and she closed and locked the door. Not long afterward she got in touch with the ever-receptive management of the Ringling Brothers and Barnum & Bailey

Circus and arranged to sell it Buddy and two chimpanzees for ten thousand dollars. Massa had gone for a mere six thousand, but he had been younger and smaller. Besides, Buddy had that ferocious expression, which was a great asset for a circus animal.

As soon as he was the Ringling Brothers' property, the circus changed his name to Gargantua. His transfer to the circus was probably not a tremendous shock to him, because Mrs. Lintz's assistant, Richard Kroener, who was also part of the deal, went with him. All through the annals of gorilla keeping, we come across close relationships between the apes and their keepers, and most knowledgeable zoo directors and circus people believe it important to observe such preferences if it is at all possible. But such softer aspects of Gargantua's personality were not permitted to leak out to the public. It was the head press agent of the circus, Roland Butler, who coined the appellation Gargantua the Great and—because people who go to circuses want to be scared—started spreading stories of his fiendish temper and incredible strength. Photographs were circulated of Gargantua apparently snarling—the poor beast's permanent expression— and colored posters were prepared to demonstrate how dangerous he was. I have a replica of one of these, showing Gargantua as big as King Kong and flourishing in one mighty fist an African native, whom he is using rather like a golf club. On the poster, in red, blue, and black print, is the legend:

RINGLING BROS. AND BARNUM & BAILEY
COMBINED SHOWS
THE LARGEST GORILLA EVER
EXHIBITED—
THE WORLD'S MOST TERRIFYING
LIVING
CREATURE!
GARGANTUA THE GREAT

What with all this change and excitement, it is hardly surprising that Gargantua's temper actually did deteriorate. Not long after joining the circus, he attacked Kroener, but the keeper managed to get away. After that the circus people began to treat Gargantua with respect—especially when he was measured and the statistics were made public. This had to be done at long distance, of course, while he remained in his air-conditioned,

49

glassed-in house. He was five feet seven and a half inches tall and had an arm reach of nine feet. Actually many gorillas are bigger, but they weren't in the circus. Years later, when he had died, it was announced that he would have taken a size eleven glove if he had ever worn gloves, and his shoes would have been twelve DDDD.

Gargantua was a tremendous attraction. According to Gene Plowden, an editor-reporter who has written a biography of him, the beast practically saved the Ringling Brothers' financial neck in 1939 and 1940. Then the management, fearing that the novelty might fade and the receipts drop off, began thinking of what novelty it could devise for 1941, and this brings us back to Toto in Havana. It seemed obvious that a bride for Gargantua would be an asset. For years, along with other circus proprietors, Ringling Brothers and Barnum & Bailey had tried to persuade Mrs. Hoyt to let it have Toto, but she had steadfastly refused even to talk the matter over. Now, however, just at the right psychological moment, she received a telegram from John Ringling North, then the president of Ringling Brothers. He said that he was coming to Havana and would like to talk about Toto. Naturally Mrs. Hoyt had heard of Gargantua and knew what must be in North's mind. She agreed to a meeting, and North came to the house. Almost at once Toto scared him nearly to death by climbing up and peering at him through a window. "Mr. North was somewhat apprehensive," Mrs. Hoyt wrote of this encounter. "But he appreciated her beauty as she hung there silently." Mrs. Hoyt reluctantly sold her baby, but on the understanding that Tomás would remain with her and that Toto would live in the same kind of luxurious surroundings that housed Gargantua. If the two gorillas didn't hit it off and if Toto were unhappy, Mrs. Hoyt would take her back to Cuba.

During the weeks that followed, while the press agents jubilantly announced the gorillas' engagement, Mrs. Hoyt kept busy preparing a trousseau—a pastime that, she wrote, saved her from "many a spell of weeping." She prepared a wardrobe of shirts, sweaters, and socks, as well as blankets, sheets, and mattress covers embroidered with the name Totito. There were plenty of bedclothes because Toto's bedding had always been changed every day. There were also Toto's special cooking utensils and a mass of food, including Cuban black beans, of which she was especially fond, and there were manicuring implements and a traveling medicine chest. At the last minute Toto, as if to show her adoptive mother how right she was to be nervous about the departure, kicked up one final magnificent shindy. It was raining hard that day, and Toto had to be locked into her playroom. In Mrs. Hoyt's words, "The baby was not in a very amiable mood." To be explicit, Toto found that one of her doors did not fit quite right, picked up a heavy mahogany sofa, and used it as a stepladder on which she could stand and batter the weak door, using her powerful shoulders. Mrs. Hoyt, fearing, as usual, that the baby would catch pneumonia in the rain, called on the whole household to help hold the door in place. It stayed shut—barely.

Mrs. Hoyt had long since decided that she would follow Toto and Tomás to America. Who else besides Tomás could make sure that the baby was getting the right food and treatment? So she flew over to Miami and was at the pier when the ship carrying Toto and Tomás docked. Toto had been awfully seasick on the way and was delighted to see her. Mrs. Hoyt then followed by car when a train carried the gorilla on to Tampa and then to Sarasota, where Gargantua was. She noted with disapproval that the newspapers carried a lot of stories about the "future Mrs. Gargantua" and that sort of thing. "It was all a little silly when the baby was actually still a nine-year-old child and Gargantua himself, after all, only an eleven- or twelve-year-old boy," she wrote.

She was there, along with about a hundred newspapermen, when the gorillas first saw each other. Toto's air-conditioned wagon was moved up close to Gargantua's, with steel doors separating the two wagons; then the doors were opened, and the two saw each other, still separated by bars and a safe distance. For a moment they stared. Then Toto barked angrily. Gargantua behaved much as one would expect an animal whose nickname is Buddy to behave: He reached out between the bars in what looked like a friendly overture. Toto stamped and barked again. He tossed her a stalk of celery; she hurled it back in his face. Gargantua retired to sit moodily in a corner, and the steel doors were shut.

In Chicago another attempt was made to let the gorillas see each other. "This time she tried to throw her bed at Gargantua," Mrs. Hoyt reported sadly. Further attempts were made now and then to acquaint the animals, but Toto was never receptive. Perhaps things would have been different if Mrs. Hoyt had left her alone, but Mrs. Hoyt didn't. For a long time she couldn't. She followed the circus wherever it went, often sitting in the cage with the baby and sometimes getting her clothes torn off in the dear old way. People loved to watch them together, and no wonder; one could never be sure what might happen. Once Mrs. Hoyt tried to leave Toto to her fate, returning alone to Havana. She couldn't bear it, and soon she was back on the circuit, following Toto around. Finally she moved to Sarasota.

Gargantua died near the end of 1949, of double pneumonia, and in a way he must have been glad to go because his teeth were in rotten shape. In her book Mrs. Lintz told a pathetic story about one of the last times she saw him. Unlike Mrs. Hoyt, she seldom paid her old pet a visit, but on one such occasion she found him asleep. When he woke up, he recognized her. He got to his feet and beat his breast in delight. Then "he crouched down again, as close to me as he could get, and with one finger pulled up his scarred lip," Mrs. Lintz wrote. "He had a gumboil over one of his last baby teeth, and he had confided his trouble to none of his keepers. But he knew I wouldn't let him have an ache in his mouth, or anywhere in his great body, if I could help it." She concluded, "Now all I could do was to report the sore tooth, and go home."

The harsh fact is that nobody dared go into the cage to take care of Gargantua. Toto wasn't allowed to go in. Tomás later insisted that after Gargantua's death she grieved for him. If she did, her native stoicism concealed her feelings from everyone but Tomás, though she could hardly have failed to notice that he was no longer there in the cage next door.

Toto herself survived nearly two more decades. She died in 1968, at thirty-six, which was a good age to have been achieved by an obese old lady of a gorilla who, shortly before she passed on, weighed in at 575 pounds. Mrs. Hoyt had her buried at Sarasota and saw to it that the grave was decorated daily with fresh flowers. Then the sorrowful woman went traveling on the Continent, as she had not dared do since she had become a widow. She was killed in Vienna in an auto accident in 1969, a scant year after the death of her baby.

In retrospect, it seems possible that Gargantua and Toto might have managed to build a life together if only someone in authority had dared take a chance and put them in the same cage. After one big scene and a lot of action all might have gone well. But nobody did dare. For one thing, the animals were worth a lot of money. Besides, there was Maria Hoyt to reckon with.

Sooner or later one who observes captive gorillas and their guardians is bound to come up against moral questions posed by members of the public.

For instance, a large number of people were deeply outraged by the behavior of Mrs. Hoyt in regard to Toto. Here was this stupid rich woman, they said, satisfying her warped maternal instinct by keeping an animal in expensive luxury instead of giving her excess money to needy human beings. Then, when Toto grew big enough to be a nuisance, the woman handed her over to a circus, there to be exposed to vulgar publicity and the endlessly gaping crowd. I am not quite sure what I think of the morality of *l'affaire Toto*, though there is no doubt that it provided fascinating information about the species *Gorilla gorilla*. What is beyond doubt is the sincerity of Mrs. Hoyt's *apologia pro vita sua* at the close of her book. Inundated with letters of criticism, she asked the world what else she might have been expected to do in the circumstances. She said, in effect, that there she and her husband were in the forest in French Equatorial Africa, staring with horror at "that terrible heap of dead gorillas," which looked even more human in death than they had in life. There was Toto in an African's grasp, kicking and biting and scratching, "her baby eyes wide with terror," and she seemed even more human than her dead relatives. And then, placed in Mrs. Hoyt's arms, "she became suddenly quiet" and nestled her head against the new warmth, clinging to the new arms, begging for protection, and trusting the woman to give it. Mrs. Hoyt adds, revealingly, that she gave no thought to the future at that moment because she was too filled with guilt. That it was her husband's finger, not hers, that had pulled the trigger for the first shot did not matter, she felt; it was *her* guilt. That was why all the rest of it happened.

Over the years Mrs. Hoyt had often remembered the words of a Paris pediatrician she had called in to attend Toto: "You have here a charming fool. She will never be anything else." In sum, the trouble was that Toto could not be treated, however kindly, as one treats a dog, yet she was not quite human either. Mrs. Hoyt reflected that it was probably like bringing up a mentally retarded child, though Toto's enormous strength rendered the analogy faulty. If I can see any moral to the Hoyt story, it is that women should not pick up baby gorillas.

4. GERTRUDE DAVIES LINTZ: STAGE MOTHER

WITHOUT INTERNAL EVIDENCE, or even with it, it is impossible to know just how old Gertrude Davies Lintz was when she married Dr. William Lintz (or, if it comes to that, how old she was, period). When she died in 1968, her husband made it known that he wanted no publicity about her, and his wishes were respected, though it is hard today to figure out why he made this stipulation. Possibly he was tired of publicity; she certainly got enough of it during her lifetime. In her book *Animals Are My Hobby* she doesn't tell us much about dates, but there is plenty of other material. She was born in Great Britain, she says, to a Welsh father and a mother who, though delicate, managed to bear twelve children before dying while attempting to produce a thirteenth. Mr. Davies had a mellifluous voice and hoped to be an opera singer before he found that his vocal cords simply didn't have enough stamina for that career. Instead, he went into the church and became a highly successful preacher. While Gertrude and about half a dozen siblings were still quite young, he took his family and moved to the United States.

Young Gertrude loved her existence in quiet country towns, mainly because farming communities had so many animals. After her mother died, when she was about eight, home was not pleasant for her, and like her brothers and sisters, she tended to stay away for longer and longer periods, sleeping rough in pleasant weather unless some neighbor took pity on her and invited her in for a night or so; the preacher was popular, and no doubt they made allowances for family eccentricity. When Gertrude was ten, Davies married again and with his wife tried to collect the children and set up a more regular home, but though the others accepted their stepmother, Gertrude did not. She still stayed away a good part of the time, accompanied during her wanderings by a stray terrier she called Rex. All this came to an end when the whole Davies family, including Gertrude, went to upstate New York for a summer engagement at a country church. There they became acquainted with a man Mrs. Lintz refers to simply as Uncle David—because, she said, he wouldn't have liked to be further identified. He seems to have lived in Philadelphia. He was very rich, liked such sports as hunting and harness racing, and interested himself in church affairs even during the summer vacations, the reason why he was in the country-side that season. Shooting, fishing, and attending church, he encountered the preacher's daughter and was charmed by her personality. Evidently he was a philanthropist who made himself responsible for the education of many worthy, needy children, and his admiration of Davies did not blind him to the fact that the preacher's children were neglected. He offered to adopt Gertrude legally, but Davies would not permit that, though he seems to have been willing enough to accept practically anything else that Uncle David offered his family. When Gertrude reached the age of sixteen, she found out, she could herself decide to be adopted by Uncle David, and in the end that was what she did, taking two siblings with her when she went to live in Philadelphia, with a housekeeper provided by him. She never saw her father again, though he lived to a hale and hearty old age.

From that time on Gertrude was free to indulge her passion for animals. She was never without a dog, she says, and a little later she acquired a gibbon named Suzette. This was her first great ape, if one accepts that gibbons are great apes—most primatologists do—and though it was not as amusing a pet as chimpanzees were to prove to be, she loved it, of course. Suzette accompanied her whenever she went out in her electric coupe. Next to animals, motoring was Gertrude's greatest enthusiasm, but she also much enjoyed harness racing, driving Uncle David's splendid horses at county fairs.

"A red-haired girl, a white horse, a blue ribbon—an inevitable combination," she wrote. To judge from the photographs, she must have been quite something before she put on weight, tall and—though we can't see that and have only her word for it—Titian-haired. The question naturally arose, What was she to do with her life? Like her father, she had a fine voice, so Uncle David sent her to Enghien-les-Bains—with a chaperone, of course—to study singing. Alas, the maestro told her that her vocal cords were "phenomenally weak," again like her father's, and she had to forget dreams of concert and opera. But she did learn French on that trip, and she made a tour, with her schoolfellows, of Italy, Spain, Germany, and Switzerland before returning to Philadelphia. Soon she crossed the Atlantic again. Uncle David had decreed she was to be "finished" in England, at a school just outside London.

By the time that stint was over, Gertrude knew that she wanted to maintain a kennel of Saint Bernards. She had seen the famous monastery in Switzerland where these dogs were trained to help struggling travelers, and her heart was set on creating a breed of these "Saints," as she called them, suitable for American house pets. She went back to Switzerland especially to buy a pair because, as she later explained, she didn't realize that dogs from the Rigi were not good as household animals and certainly not ideal companions for children, being inclined to suspicions and savagery.

She attended her first dog show in 1909, in Long Branch, New Jersey, where it was sponsored by what was still known simply as the Kennel Club. Her entrance with the dog Rigi, as she describes it, was spectacular. She was in white, and the dog wore a muzzle, which was against Kennel Club rules; but Gertrude didn't know that. "The muzzled brute on a leash, and the Titian blonde girl behind him," she wrote, "and in their lee a maid carrying my parasol and rubbers . . . caused a sensation, partly of course because the dog should not have been leashed or muzzled, but partly also because it was so striking a sight." All the papers, she recorded proudly, carried the same headline over their stories: BEAUTY AND THE BEAST. To be sure, she was immediately asked to leave, because of the muzzle and the leash, and of course, she did, one supposes in maidenly confusion, but at least she now knew the ropes. At any rate Gertrude Davies with her dogs was admitted to American Kennel Club shows as soon as it became established as arbiter of these matters, so she could not have offended the members. Strength and beauty were her aims, she maintained. She thought of Hercules and Venus and made of the two names her own kennels' name, Hercuveen. At the Westminster Kennel Club Show at Madison Square Garden, New York City, in February 1910, her dogs won their first ribbons.

It was a happy time. Kennel magazines referred to her as the Joan of Arc of the American Saint Bernard. For twenty-five years she showed her dogs herself. It was, she said, a strenuous life, and because of it, she developed powerful muscles. Unlike her delicate little mother, she was of large build, a strapping Joan indeed.

Unfortunately this success story was interrupted when, after three years of breeding and showing, the Hercuveen dogs were smitten with a mysterious disease that killed off nearly all of them. Gertrude herself became gravely ill from the strain and had to go into the hospital. There she met Dr. William Lintz, who was teaching bacteriology at the Long Island College of Medicine, and in April 1914 they were married. Though Lintz wanted to build a practice, he had done little toward his ambition apart from lending his services to the hospital. Now he dared cut off all his activities except for private patients, who in the course of time were to become very private indeed, and very special. It is not every doctor who finds himself ministering to ailing anthropoids.

Gertrude and her husband moved to Shore Road in Brooklyn, to a large house with a two-acre backyard. This time she kept only thirty or forty Saint Bernards, and it was possible to house them all, along with any other animal that caught her interest. For a time she bred fancy rabbits and tumbler pigeons as well as gave houseroom to a huge lizard that might have been a komodo (at least she referred to it more than once as a Chinese dragon). Following theories of her own, she fed her dogs meat containing thymus gland, which she got from the butcher from among his castoffs. The thymus, she said, had the effect of making the animals grow very large. By the third generation some of them reached a weight of 250 pounds and stood more than three feet tall at the shoulder, but apart from being splendid to look at, such animals were not really practical as pets. They were too heavy for their hindquarters, were vulnerable to various diseases, and their puppies were sometimes so large that they had to be delivered by cesarean section. After a while, therefore, Gertrude gave up the thymus feedings but otherwise retained her special diet. One of the first breeders to understand the importance of vitamins, she fed her dogs brewer's yeast, wheat germ, and fresh liver. Soon, however, she suspected that wheat germ increased a tendency to cancer, and she stopped using it.

What with the rabbits, the tumbler pigeons, other breeds of pigeon, hummingbirds, giant horned owls, and even, for a while, a leopard, it must have been a very busy backyard. One wonders how she kept all these creatures safe from each other, but she had two men to help who, during the time they worked for her, became, perforce, expert naturalists. They were there, fortunately, when Purser, the leopard, took exception to the scent of the dogs on Gertrude's clothing and attacked her. After that Purser was sent to the local zoo.

As for the great apes, at first it did not occur to her to keep any. They were not, to say the least, anything like so fashionable a hobby as Saint Bernards, nor were they as readily available. Though many zoos owned chimpanzees and one or two even exhibited a gorilla or two, the anthropoid apes had not been studied very carefully, and not a lot was known about them. Gertrude saw her first chimpanzees during her honeymoon, when she and Dr. Lintz visited the Vienna Zoo. They were a pair, and she was favorably impressed by the behavior of the male when some boys teased them. He drove them off with a display of anger, then went to his mate and put his arms around her as if comfortingly. Mrs. Lintz reflected that he was a nice animal. Later, back in America, she saw at the Ringling Brothers' Circus a baby gorilla. It was ailing, as were most captive gorillas in those days, and she yearned to tend it. She knew, she wrote later, that she could cure it if only she had the chance. Day after day she returned to the circus just to look at the infant, but she didn't quite like to volunteer her services; who was she to step up and claim special talents? So she went her way, and the infant inevitably died. Later, having met a Ringling, she told him about her reaction to the animal, and he remembered it and said that he had wanted to ask some woman—any woman—to help nurse the little animal.

"But there were too many taboos involved," he said.

Gertrude's first chimp was a young female named, mysteriously, Maggie Klein. Why Klein? Perhaps it was the name of the person from whom she acquired the animal. Mrs. Lintz deliberately treated Maggie just as she would a human child. The little ape learned to brush her teeth and go to the toilet; she even had a daily bath and enjoyed it, though chimpanzees are usually afraid to immerse themselves in water. Maggie knew how to soap herself, and did. The only drawback, according to Gertrude's ideas, was that the chimp didn't develop independence as soon as the dog breeder thought she should. She depended on Gertrude emotionally to such an extent that she was a nuisance, clinging to her adoptive mother and wailing whenever she was left alone. In fact, all chimpanzees hate to be alone; Mrs. Lintz found this out the hard way. Very well, she decided; she must find Maggie a companion, preferably one of her own species.

Then began the story, told in the previous chapter, of Gargantua. We have been told how Gertrude asked their friend Captain Phillips to bring her a gorilla. Phillips said he would try, but before he got back, the same friend who had supplied Maggie Klein brought a new young chimp to

Shore Road—Joe Mendl (again, why Mendl?), one year old to Maggie's two years and somewhat smaller than the little female. He was welcomed, of course, and immediately began to follow Maggie much as she followed Gertrude. But as the latter had hoped, Maggie's overfond attentions abated as she concentrated on the little newcomer. Mrs. Lintz learned a lot about chimpanzees from this couple. For one thing, she reported, when amused, they could laugh aloud. Nor was she the only person in the world to be fascinated by chimpanzees. Joe enjoyed hearing the sound of his own footsteps, so she took him to buy a pair of shoes. They went to a Fifth Avenue shoe store and were mobbed, indoors and out.

The *West Key Bar* returned to New York in 1928, bringing three little chimps for Gertrude, but no gorilla. All the little apes were very poorly. She listed some of the ailments that were so easily picked up from human captors in West Africa: pellagra, hookworm, intestinal worms, and, often, tuberculosis. Mrs. Lintz accepted the condition of her new charges as a challenge and quickly got to work feeding them milk, eggs, liver, iron, cod-liver oil, and vitamins. Even so, she admitted, they "caught everything going" all the time. When Maggie was six, and Joe five, they caught first measles and then mumps. It made them very sick, and when, ultimately, Gertrude started exhibiting her apes, she took measures to protect them from being infected by the public, keeping them in what she called germproof cages that were made of closely fitted plate glass.

The chimps were very clothes-conscious and understood the words for various garments. Maggie Klein actually tried, after watching her mistress sew, to thread a needle. She tried against great odds, because her thumb was too short to oppose her forefinger. Her attention span, too, was short, but she was stubborn. She would try to thread the needle, fail, tire of the game, go and do something else, and then come back to the needle and try again. One day, after five months of such sporadic attempts, she actually succeeded. After that, always with one eye on Mrs. Lintz, who liked to sew and did a great deal of it, Maggie started to do that, too. She would hold her cloth with one hand and stab it with the needle held in the other, turn the cloth over, and pull the needle through, then go through the action again from the other side. She really did sew. As for Joe, he was a mechanical genius. He could saw, plane, hammer nails, and take them out again either with the clawhammer or his strong little fingers. He was very clever, too, with keys, and if he got the chance, he would unlock his cage door and all the other cages as well. Often he unscrewed Maggie's exercise trapeze so that she fell off, and he also snatched away, whenever he got the chance, the cloth she carried everywhere. We would call it her security blanket, but the Lintzes referred to it as Maggie's prayer rug. In revenge she held water in her mouth until she got the chance to spit it all over Joe. He hated to get wet, and she knew it.

Gertrude went further and further into her training of the apes. Because, as she found out, they had a strong sense of rhythm, she taught them to shimmy and to play the harmonica. At least Joe played such tunes as "Turkey in the Straw" and "Home Sweet Home." The tones were not particularly accurate, Gertrude admitted, but without too much imagination anyone could recognize them. Maggie, for her part, did a very convincing striptease. At last Mrs. Lintz decided that so much talent should not be confined any longer to 8365 Shore Road, Brooklyn; it was time to take the show on the road. After all, hadn't she been showing her Saints for years? She enlisted the services of a professional animal trainer, and with his help she took Maggie and Joe onto the stage, into vaudeville, where for two years, she proudly said, they appeared on the better circuits in New York. She noted that they were born troupers and loved applause. It is an observation invariably made by those with a more than casual acquaintance with chimpanzees.

From vaudeville it was a natural step to the Chicago Century of Progress Exposition in 1933 and 1934. The fascination of footlights now had Gertrude, as well as Maggie and Joe, firmly in its thrall, and it was in Chicago that Bob and Mae Noell, who will be described in due course, saw the group, which they remember vividly.

"Mrs. Lintz was in the cage," said Mae Noell, "wearing a jungle suit— sort of khaki-colored, you know, with a bush jacket and a white pith helmet. One little chimp was the star, as I remember. Mrs. Lintz handed him a tin top of some kind, like a saucepan lid or something like that, and a hammer and a nail and a stick. Well, that little chimp took the hammer and pounded the nail into the end of the stick. Then he took the tin top and put it on the nail and carried the whole thing around the stage—the cleverest, cutest thing you ever saw. When the tin top fell off, he put it back. That's when Bob said we'd simply have to get a chimp sometime." That must have been Joe, the mechanical genius.

The Shore Road menagerie was branching out. Captain Phillips did at last find a gorilla and brought it to Brooklyn, along with his customary (and ever-growing) group of chimpanzees. To tell the truth, it wasn't much of a gorilla, being, as Mrs. Lintz said, a pitiful little handful of skin and hair in a grocery carton, but it was her first gorilla, even though it seemed to be dying of pneumonia. She thought, mistakenly as it turned out, that it was a member of the rare mountain species (the Martin Johnsons had been busily educating the public on the difference between lowland and mountain gorillas). Its plight reminded Gertrude of the forlorn little gorilla at Ringling's and smote her heart. She worked hard over it, and in the end the little creature recovered. The disease had a strange effect. Nearly all of the gorilla's hair came out during the worst of the siege, and now when it grew back, much of it came in white. Mixed with the natural black hair, it looked silver. As the world was to find out, male gorillas reaching maturity grow black-and-white hair like that, for which reason they are called silverbacks. But Massa, named after the Massa in southern dialect, as in "Massa's in the cold, cold ground," was too young to be a silverback and, moreover, was believed to be female. It is very hard to tell which is which with immature gorillas, but Captain Phillips thought it was a female. Gertrude accepted his opinion and continued so to do. For the next four years, which he spent in Brooklyn, Massa was considered a girl gorilla. Ultimately the mistake was discovered, though not before he was transferred to his lifetime locale, the zoo in Philadelphia. In the meantime and even after she had learned better, Mrs. Lintz could not stop thinking of him and writing of him as a female—which might account for Massa's occasional shortness of temper. For example, his mistress noted approvingly that Massa was a good little mother when she played with the infant chimps and excelled at imitating housewifely arts such as scrubbing and cleaning.

There was a strong bond, at any rate, between woman and gorilla. For a long time Massa, still convalescent, would not accept food from any hand but Gertrude's, and Gertrude noted that she herself was beginning to understand gorilla language or, at any rate, Massa's language.

"She has a word, full of disgust and rebellion, for her cod liver oil," she noted in *Animals Are My Hobby*. "Food she liked was welcomed with a low gurgle." She also commented that the ape's nervous system was so nearly human that if she became emotionally upset, she fell victim to acidosis. This should interest the student of Dr. Yerkes's work, for he made exactly the same discovery about Congo, the female gorilla he observed in Florida for several seasons between his university commitments. One day, when Congo resisted being weighed in the scale and he persisted in placing her there, he had no idea that she was as upset as she was, because she showed none of her feelings facially or vocally. She just suddenly had diarrhea. From this instance Yerkes concluded that gorillas are introverted rather than extroverted like chimpanzees. In a similar situation a chimp would have screamed hysterically and resisted with all its might.

On reading the Yerkeses' book *The Great Apes*, Gertrude in 1932 was much impressed and wrote a fan letter to Dr. Yerkes. He replied promptly, as he always did when he heard from people who owned gorillas, asking her many questions about her rare specimen, and a flourishing correspondence ensued. She sent him photographs of the Lintz apes and a long panorama picture of the Hercuveen Saints, standing shoulder to shoulder, most impressive. If the truth were told, however, she was rather losing interest in her dogs, young gorillas and chimps giving her little leisure for anything less human.

In one experiment she gave a kiddy car to Massa and another to one of the chimps. The chimp punished his kiddy car, banging it around and quickly breaking and forsaking it. But Massa experimented with the toy for a little, learned to pull it along the floor, and after several days was propelling himself in it. The chimpanzee, Gertrude decided, uses trial and error to reach his conclusions, but the gorilla thinks things through. The gorilla is constructive and gentle, words that cannot by any stretch of the imagination be applied to the chimpanzee. Besides, Massa was the most affectionate of all the Lintz apes. She (it is hard to get away from Gertrude's pronouns) tried on all her mistress's clothes, powdered her face, and adjusted her hat—no wonder Mrs. Lintz persisted in considering her a female, even after she knew better.

But now another actor was added to the Lintz cast of characters. The story of Buddy, or Buddha, or Gargantua has been told in the preceding chapter about Maria Hoyt and Toto.

Under Gertrude's watchful eye both young gorillas improved in health and strength, but she was alert, as she was when observing her Saints, to dietary needs and preferences. It seemed to her that they liked protein foods and searched them out, even though gorillas were believed to be completely vegetarian. (For one thing, they ate worms and insects when they got the chance.) In cases of doubt she had always fallen back on liver, and now she prepared strips of liver lightly boiled so that they looked like worms; the little apes ate them eagerly. Encouraged, she moved them on to minced beef, and this, too, they liked. But Massa, though there was nothing wrong with his digestion, in time fell ill with a complaint Mrs. Lintz described as something like polio and became paralyzed. In fact, very probably it *was* polio, though not until another little gorilla at the Yerkes Primate Center, years later, evinced similar symptoms did the scientific world agree that the great apes could and sometimes did catch the disease. For Massa's paralysis Mrs. Lintz rigged up an arrangement of ropes and pulleys and showed him how to exercise on them. After some weeks of this treatment the stiffness disappeared.

Gertrude commented more than once that gorillas are deadpan artists, but Buddy was unable to hide his emotions when he felt fear. Both ape children loved motoring as much as their mistress did, but Buddy never seemed confident that it was quite safe. He sat beside the driver, one hand on her arm, the other on the steering wheel, all the time they were in motion, and though he never flinched at the approach of ordinary cars going in the opposite direction, the sight of a big truck or streetcar coming along made him panic. He would yell at the top of his lungs and cover his eyes with his hands or bury his head in her skirt, which must have been disconcerting for Gertrude until she got used to it. Massa, however, never seemed to fear any part of the traffic. The great Saint Bernards intimidated Buddy for a time until he got used to them, and he even showed fondness for the dogs; but he never became reconciled to seeing Dr. Lintz's saddle horse, which was brought to the front door every morning. He was just naturally afraid of big things.

Though young gorillas usually walk upright before their increasing weight makes it more convenient to drop down to a four-footed, or four-handed, position, Buddy seems to have preferred his four feet (or hands) on the ground all the time. Gertrude coaxed him to stand up with a strata-gem. He loved fruit, so she would put a piece of fruit, a banana or orange, in each of his armpits and then fill his hands with peanuts. To save the fruit, he simply had to stand erect until he had eaten the peanuts, and little by little he got the idea. No back-to-nature theorist, Gertrude put all her other apes in sneakers because, as she said, they could feel the ground under them that way, but as we have said, Buddy liked to make a clumping sound, so she had steel plates put on the heels of his leather shoes.

When eating, he made noises of appreciation—grunting, gurgling, and smacking his lips. For the benefit of anybody else who might wish to rear a delicate gorilla, Mrs. Lintz went into detail about his meals. After a year's stay on Shore Road he was swallowing, every day, a quart of milk with a raw egg beaten into it, sweetened with two tablespoons of corn and chocolate syrup; half a pound of parboiled liver; six average-size bananas; three or four apples; one head of lettuce; several stalks of celery; some sweet potatoes; a bowl of cooked rice or oatmeal with sugar; cakes; and raisins. He appreciated extra treats such as ice-cream cones and lollipops. He was very particular about his food; apples could not be bruised or blemished or he would discard them.

When the Chicago Century of Progress Exposition started up in 1933, she took her apes to the fair and proudly exhibited them. The performing chimpanzees went along, and both gorillas. (The Noells don't seem to have noticed the gorillas at the time probably because they were so wrapped up in chimpanzees.) Mrs. Lintz brought along her two regular assistants as well as the trainer who had chaperoned the vaudeville acts. On the fairgrounds the troupe was quartered next door to Sally Rand, the famous fan dancer, and very soon Maggie Klein was doing a burlesque of the Rand fan dance. In the striptease she was wonderful, said Gertrude. As she removed each garment, she carefully smoothed it out before turning back to her stripping. Buddy, though not as much a show-off as the chimps, was not above joining in the act and gaining his share of applause. He loved to pull an empty sack over his head and body, then do a blindman's buff performance. One day, in view of the audience in their glass, germproof cage, he was playing with a kitten, and when Mrs. Lintz tried to get it away from him, he would not let go. To make him do so, she bit him. Buddy got the message and released his pet, but as a spectator said later, "Lady, when a woman bites a gorilla, *that's news!*"

Show biz did not come to an end when the party returned to Shore Road because Mrs. Lintz had a new idea. If the general public was so fascinated by her apes at the world's fair, why shouldn't they care to look at them at home? So a "Simian House" was constructed in the backyard, and the animals were put through their daily routine for public delectation— brushing their teeth, dressing, folding their blankets, and taking lessons at table in how to eat politely. Of course, the apes were not always letter-perfect in their demeanor. To keep them in hand, Gertrude and her assistants used a certain frightening mask obtained somehow from the

South Seas. This always scared the gorillas into being reasonable, but then they were timid about many things—thunderstorms, rats, mice, and snakes, most of which, we might note, frighten a lot of humans as well. Buddy, as we have seen, was afraid of many things that didn't frighten Massa at all. They both were playful, however, and their games were like those of human children, such pastimes as hide-and-seek, tag, and ring-around-a-rosy. They enjoyed swings, teeter-totters, trapezes, and horizontal bars. It was a good thing, reflected Gertrude, that they had each other to play with, because a solitary ape is a man-made freak. Oddly enough, they did not mind children, though the chimpanzees hated them. Children, the chimps seemed to feel, diverted attention from themselves.

For years Gertrude corresponded with Madam Rosalía Abreu, as did practically every ape fancier. Often, of course, she wrote to Dr. Yerkes, too. She added to her collection by procuring an orangutan, but he was never really healthy and always had a cough, so he wasn't much fun. In her book she mentions that Massa could and did on occasion weep with real tears, but one doubts this. Buddy would wail, but tearlessly, when he was lonely and once when his hands and feet were cold. Massa actually permitted Gertrude to take food away from her, though not *favorite* food. Buddy, always affectionate, had been known to offer his mistress a lollipop or at least a lick of one. Maggie, however, never permitted such liberties.

It sounds an idyllic existence for a woman who loved apes, but the idyll came to a jarring end when Massa attacked his mistress (as described in the preceding chapter). Blood flowed. Being allergic to antitetanus serum, Gertrude had to spend the next three days in a tub of Epsom salts solution. She had twenty-two wounds but refused to go into the hospital for fear word would get around and some official would insist on destroying Massa. One night, when she was somewhat better, she went to see the gorilla. Massa seemed embarrassed, she said, and avoided looking at her many bandages. To test her dominance, Gertrude gave a trivial order, telling Massa to sit on a certain chair. The gorilla would not obey; in fact, when she gave another command, he barked in defiance. Mrs. Lintz went back to bed.

She did not decide at once to end the relationship. She gave herself time and on another occasion went to Massa's part of the estate to see the gorilla. All went smoothly until she started to go, and Massa prevented her; the assistant Dick Koerner had to come to her rescue. On another occasion Massa pushed her down. This settled matters. Massa had to be caged, at least, while Gertrude thought things over. She wrote to Dr. Yerkes asking what she should do about selling Massa—she didn't mention the recent unpleasantness—and Yerkes was shocked. Sell her gorilla? It would be a shame, he said, to abandon the experiment at this point, when she had gone so far and observed her rare pet so carefully. Finding her determined, he wrote that his experimental center could not possibly afford Massa, but he was sure that almost any good big zoo, especially Philadelphia, would be delighted to offer the animal cage room, and a good price as well.

Thus in 1935, just before Christmas, Massa and Gertrude took their last motoring trip together and drove to Philadelphia, where all had been arranged. Sadly Mrs. Lintz handed over the gorilla, weeping with pride, she tells us, as scientists gathered to admire the beautiful silver-haired ape. Some years later those same scientists concluded that Massa was, after all, a male. In the meantime, the animal's former owner stayed in Philadelphia for three weeks after the installation, to help him settle down. Massa remained well and truly settled, and at the moment of writing he holds the record for longevity among his captured species—fifty-five on his last birthday. Belatedly he did Gertrude proud, dying at the end of 1984.

At this point the Lintzes were going to Florida every year to spend the winter, of course accompanied by the entire household of apes, but Gertrude had faced reality and given up the Hercuveen Kennels in order to concentrate on the primates. She acquired another gorilla named Skippy, which soon died of a blood infection. Gertrude had so many chimpanzees now that she kept distributing them to shows. One, named Jacky, went to Hollywood and appeared with Johnny Weissmuller in the first Tarzan picture. Ringling got Jacky Two. Jacky Three found a berth in another Tarzan picture. Maggie and Joe continued to tour in vaudeville, coming home between seasons to rest. For a long time the Lintzes gave houseroom to another talented chimp, Jiggs, but she—for she was a she—didn't go professional; she stayed at home with the old folks. Sometimes the assistants, Tony and Dick, took the main troupe to Coney Island to appear onstage, and everybody, apes and people together, spent one summer at Atlantic City. For the World's Fair at Flushing in 1939 and 1940 Frank Buck borrowed a few of Gertrude's chimps. She herself took some of them to a smaller fair in Toronto in 1944, and they all were stranded by bad weather at Niagara Falls. In other words, it was just an ordinary existence for show business people, and it was fun.

But Massa's departure had left a vacuum in Buddy's life, and his temper began to get uncertain. For one thing, he had a new, disturbing trick of biting people's fingers. At six he was strong enough to worry his enemies if he wanted to. The year after Massa went away he had another tragic experience with acid in Florida when a juvenile delinquent fed him poisoned chocolate syrup containing disinfectant. Two chimpanzees that also got the syrup died. Buddy was very, very sick, but he survived, though in the process he lost eighty pounds. The experience left him with an increased distrust of men and an even more soured disposition. He escaped one night for several hours, disporting himself amid the thick jungly vegetation of Florida. He was recovered without incident or publicity, but Gertrude was afraid it might happen again. And he might easily have attacked the hunters who came after him that time, she reflected.

Then came the incident of the storm and Buddy's joining her in bed. Mrs. Lintz faced facts. It was only a matter of time, she knew, before another crisis would occur. The gorilla had to go. He was a famous animal and a prize; any zoo would have been delighted to take him. But Gertrude was aware that circuses, too, wanted gorillas, and she decided on the biggest one in America, Ringling's. It was now being managed by the original founder's nephew John Ringling North, and she got in touch with him. Of course, he jumped at her offer, and she drove as hard a bargain as she could for Buddy's sake, making many stipulations. As has been said, Dick Koerner had to go along with him as part of the deal, tending him as long as he was needed, though in fact, it was Dick who died, long before Buddy. Buddy at this time, about six and a half years old, weighed 468 pounds, nearly the top amount for an animal his age. His main cage, Mrs. Lintz said, had to be at least twenty by seven feet, with a smaller sleeping compartment. This, too, was agreed to. The cages had double-thick glass walls reinforced by chilled steel bars set close together. An air-conditioner maintained a steady temperature of ninety-six degrees and fifty percent humidity.

As we know, Buddy went on to become Gargantua the Great, his scarred face and evil sneer adding immensely to the effect that Ringling's wanted for its star. People would come to stare at him and shudder in delicious terror. By extension, Gertrude, too, became famous, though in a much smaller way. She was interviewed by the president's wife. She wrote for the Hobby Lobby, whatever that was. She was admitted to the Order of Adventurers, the second woman ever to be so honored, and the Adventurers, whose members included Roy Chapman Andrews, Lowell Thomas, Admiral Richard E. Byrd, and Theodore Roosevelt, gave her a gold medal.

It was 1937 when Buddy was metamorphosed into Gargantua and went off with the circus. He died on November 25, 1949, at the age of nineteen. The autopsy revealed that he had died of double pneumonia and that four of his teeth were abscessed.

In time, Gertrude got another gorilla and sold it to Dr. Yerkes. Looking back over her life, in her book *Apes on Stage* she mused: ". . . my apes were trying to become something else. I cannot express it differently. With a deep and almost tragic desperation, with a will that is not in other animals, they were striving. . . ."

5. *Mae Noell: Carnival Queen*

"I was born in Charlotte, N.C., at midnight July 8/9, and always celebrated it on the 9th," wrote Mae Noell, whose name in childhood was Anna Mae Roach. "Mama said it was the 9th, and since she was there I believe it." The year was 1914. She wrote this to the author in the beautiful copperplate handwriting that is characteristic. She was a bright child; by the time she was three she had learned from her mother to write her own name. Few ordinary children could do this, but hers was no ordinary childhood.

"Aged ten days to three years: riding trains, autos and mostly horse-drawn vehicles from town to town since parents were vaudevillians," her letter continued. "At age 3 made my stage debut singing 'I'm forever blowing bubbles.' That's the year, too, that my brother was born—1917."

At five or six she was acting small parts on the stage, and she had started her education under her mother's tutelage on the road. Now and then, when the Roaches visited relatives in New York, the children went to school in the Bronx or Manhattan; at most, Mae got two and a half formal terms under her belt. For the rest of the time they were on the road. The mother and the children had a little act, dancing and singing, with their father the general manager of the troupe. Mae was twelve when he decided on a change and opened his own medicine show, but the traveling continued. Mae, her mother, and her brother, as soon as he was big enough, would dance and sing for an hour and a half, changing the program every night for a week and then starting over again; at the end of the performance Father Roach went into his spiel and sold medicine. He was good at it. The most popular nostrum was a strong vermifuge, but he also dispensed tonic, liniment, tooth powder, anything, evidently, that the source company supplied, with one reservation: There must be no alcohol in the concoctions whatever they might be, for the Roaches were strictly teetotal, as Mae still is.

"The worm medicine was the most popular," Mae told the author. "You'd be surprised how many people used to have tapeworm, and our stuff really worked. My dad encouraged people to bring their tapeworms, bottled in preservative, to the show; we'd give a little prize for the first one to arrive. Our liniment was popular, too, because its base was benzene, and you know how strong that is. Dad had a lot of special inducements for people to buy. Sometimes we'd have an evening, just before we pulled out, when you could buy a prize package of assorted medicine for a bargain price. We also sold little packets of candy, and every tenth packet would have a prize in it, sort of like Cracker Jack. Mama and I would make them up, but it was different; the prize in our candy entitled the purchaser to that bargain package. There was an extra special, too, a five-dollar package of something unnamed, the implication being that it was an aphrodisiac. Now and then Dad would slip one of those packages into the big bargain packet.

"For the show Mama and I dressed alike, with big ribbon bows in our hair. As soon as my brother was old enough, we brought him into the act, dressed up like us, except that he couldn't wear girls' shoes. So there he would be in his little starched dress and hair ribbon just like Mama and me, and we'd do a little dance that ended with a kick. And up would come his foot in a scuffed boy's shoe, and it always brought down the house."

Life on the road could not be conducive to keeping household pets, but the children were taught by their parents to be very kind to animals, as well as older people, children, and the handicapped. Animals, as the elder Roaches explained, can't tell you what is wrong, what they want, or where they hurt.

Mae was seventeen when she fell in love with Robert Marshall Noell, two years older. Bob, too, had been born in the South, in Bedford County, Virginia. He was twelve when he left his widowed father and joined a medicine show run by one Doc Elting; at fifteen he left Elting to set up on his own. Mae's father did not take to the enterprising young medicine man; but Mae's mind was made up, and the two eloped. It was a congenial match.

"We knew the same routines, so it was easy. The show went right on," said Mae.

They had been trouping for a year or so, selling medicine in the time-honored way, when in Chicago, at the Century of Progress Exposition, 1933–34, they saw their first chimpanzee. It belonged to, and was being exhibited by, Gertrude Lintz of Brooklyn. It was young and, in the manner of its kind, very appealing. Bob said, "I want one of those things!"

The germ had been planted, but nothing came of it quite yet. Six years later, in 1939, at the New York World's Fair, the Noells, who were, naturally, inveterate fairgoers, witnessed an elaborate chimpanzee act with many animals, staged by Reuben Castang.

"That was the clincher," said Mae. They watched wistfully as the apes rode bicycles, skated, danced, and otherwise did their marvelous tricks. Bob was more than ever determined to own a chimpanzee, though every time he expressed this desire Mae, now the mother of a daughter and a son, replied, "Oh, no, you're not! That thing might kill my kids!"

In fact, a year earlier they had revamped their show, giving up the medicine and acquiring instead a little traveling menagerie billed as "Noell's Ark." The public now bought tickets instead of medicine and filed in to look at the Noells' white rats, African green monkeys, and such small deer.

In February 1940, with visions of chimps still dancing in Bob's head, the family happened to be taking in the Mardi Gras festivities in New Orleans, strolling along St. Charles Street, when they came upon one of the last store shows still in existence. A store show, Mae explained in her book *Gorilla Show* (privately printed and published in 1979), was a dime museum under another name—an empty store building rented out to a group of show people who decorated it for the short time needed, set up a ticket box at the entrance, and put on a number of small exhibitions inside. This one was what is called a "ten in one," with ten platforms each with its own act. As in other store shows, a lecturer walked from platform to platform, or rather from stage to stage, explaining the attractions, now and then joining in an act and, later, selling picture postcards. It was there that the Noells saw Snookie, a little (forty-nine-pound) chimpanzee, with which Bob, of course, immediately fell in love, wearing a striped T-shirt and overalls. A four-year-old boy shared his act with Snookie. They pulled a toy wagon, taking turns to ride in it, and, also in turn, rode a miniature bicycle around the stage. Bob was completely enraptured and asked the proprietor what an animal like that might cost. The answer, three hundred dollars, seemed and was an immense sum; as Mae explains, times were hard in 1940 for show people as for everyone else.

But Bob was obsessed. When Mardi Gras was over, Mae could not drag him away from New Orleans's vicinity, though as they both knew, business was bound to be slow in Lent, in a Catholic country. Ordinarily the Noells would now have taken their little outfit north. Instead, they moved in a circle around the outside of New Orleans, always within reach of Snookie in the store show, so that Bob could go in every day to see him and to beg the owner to sell him. Luck was with Bob, or, as Mae would have said, against her, because the Noells did have just three hundred dollars in hand, borrowed expressly to buy a new car with which to tow their trailer. When at last Snookie's owner relented and said that Bob could have him, the new car went glimmering and the chimp became Noell property.

Happily Bob and the Noells' young helper went up to New Orleans to collect the ape. The owner's wife, weeping, packed the chimpanzee's clothes and other belongings, including a chamber pot which he was always to use like a well-trained child. There was also a length of rubber hose the owner threw in as a goodwill gesture. He warned Bob that it would be a good idea to keep the hose handy in case he needed it for discipline, but Bob paid little attention; he was too happy. With Snookie packed in comfortably between himself and the assistant, Frisco, they drove off toward the Ark, eighty miles south.

It was an easy trip, though it was educational. A chimp the size of Snookie can raise considerable hell, and Snookie did. He soon got hold of the rubber hose and beat both men until they were covered with bruises. But at last they reached the Ark's camp, where Snookie was introduced around the family and then put to bed. Fortunately, though the Noells were incredibly ignorant of the harm he could have done, he didn't hurt either of the children; but the baby girl, Velda Mae, inspired jealousy in him and had to be guarded thereafter. Snookie got on immediately, however, with Velda Mae's elder brother, Bobby. The Noells at last had their ape. Snookie's cold was a worry, but he soon recovered.

Even Mae could not deny that he was an immense attraction for the public. Wherever they went the show's visitors and, of course, the press made an immense fuss over Snookie. Almost everything he did was news. One popular paper told how he escaped and went into a drugstore, sending the regular patrons scattering in dismay from their stools at the soda fountain. Snookie took possession of a stool and pointed to some edible, and the clerk immediately gave it to him. Snookie ate it and pointed to something else and got that with just as much alacrity. This continued until he had rolled up a large bill, which Bob later paid without so much as a murmur of complaint—something he would never have done, his wife complained, if it had been either of his other children who had behaved like that. But then, neither of the other children would have had the clout to do it.

As Snookie's fame grew and spread, the lure of the Ark's other animals faded; nobody paid any attention to them, until the Noells faced facts and gave them up. Snookie was their only attraction, and he was enough. The Ark's name was changed and became Noell's Ark and Gorilla Show, not, as Mae explained, that they had any intention to deceive, but simply because people insisted on calling Snookie a gorilla. Perhaps, she guessed, they had trouble pronouncing "chimpanzee." As the weeks went on, Mae lost her early distrust of Snookie. The Noells learned a good deal about the nature of a chimpanzee, and perhaps Snookie, too, learned something of the Noells. Naturally he got out of his cage whenever he had the chance, and naturally, too, he got that chance more often than not when Bob was away. Once when this happened and Mae, alone, was terrified not so much of what the chimp might do as what the trigger-happy public might do to him, she lured him into her kitchen and fed him goodies until Bob got home. After they had acquired more apes, she used the same methods whenever there was an escape. The animals learned to go straight to her kitchen when they got out because there was a perpetual party going on there. So the odd little community had no more heart-stopping escapes, nor danger from the public or the police.

For a long time Snookie's mere presence was enough for the ticket-buying public, but there came a time when just looking at a chimpanzee, even if he was dressed in a T-shirt and rode a bicycle, did not satisfy. Children, of course, clamored to get closer, to pat the chimp's head and play with him, but children can usually be managed. Not so adults who wanted to stir up trouble. One man in particular, a drunk who watched Bob's struggles one evening when Snookie didn't want to go into his box, jeered and said *he* could handle the little brat if Bob would give him a chance. He could do it, he said, with one arm tied behind his back. Bob was goaded into saying, "Oh, yeah? You try. If you can do it, I'll give you five dollars."

Thus was born the Athletic Ape Show, though nobody knew it at the time. The drunk accepted the terms, so Bob sent the rest of the house out, telling them they could come in again on payment of fifty cents a head. They all came back and watched while the challenger tried in vain, for a long time, to wrestle that little chimp back into his box. Everybody except the drunk enjoyed the spectacle, so a similar stunt was tried the next day, and the next.

For a long time the rules remained much as they had started out: five dollars to anyone who could put the chimp into the box. It was amazing how many men took up the challenge and how many people of both sexes paid to watch the ensuing struggle. Snookie never tired of the game; but he was not beyond working in improvements, and one evening he brought in the unwelcome addition of running playfully into and out of the box. This was not so good; it enabled the challenger to argue that he had succeeded. The Noells had to think of a new angle, and they did. The contender, they ruled, had to pin Snookie's shoulders down on the canvas and hold them for a stated interval before he could claim to have won. All went well until Snookie impishly permitted a man to hold down his shoulders, and the rules had to be changed again. This time the contender must sit on Snookie's stomach before he could be proclaimed winner, and that did it; nobody conquered Snookie after that.

However, it is the nature of a chimp as he grows to keep testing his strength, and there came a time when Snookie was punishing Bob rather than the other way about; he bit and scratched. With some initial difficulty the Noells designed and had made a muzzle that fitted his face, and later they introduced boxing gloves into the battling chimp's costume. After this, things were not so risky, and the so-called Boxing Chimp act became the chief feature of the Gorilla Show, especially as the Noells acquired more chimpanzees, so that it was not always Snookie who did the fighting. For thirty-one years the Noells traveled around with their apes, and very seldom did they have to hand out the famous five dollars. Mae, who had a talent for drawing, designed their posters. The heading of a typical playbill was

WE ARE LAUGHING AT
THE FUNNIEST SHOW ON EARTH

and it continued:

Noell's Ark
Gorilla Show Inc.
Each Night, after the big
FREE SHOW
you will laugh, too, when you see
Several Local Men
Wrestle One of These
Big Animals at One Time.

The show developed. The Noells owned a trailer, with proscenium and arena, also living quarters for the animals as well as the family. They awarded certificates for certain contenders in reward for "good sportsmanship," for having stayed the course, for having obeyed orders, for practically anything; these certificates went to any contender who didn't break the rules as laid down by the Noells. People loved the certificates and went to a lot of trouble to get them.

It was not always smooth sailing because working with chimpanzees has its hazards. Once, for example, Bob was careless enough to wear a tie, and Snookie, in a rage, nearly throttled him with it. Bob called to Mae for help, and she arrived in time to talk baby talk to Snookie, which calmed him down. It always calmed him, she said, when she said in a singsong voice, "Sit down, Snookie, sit down."

"But it enraged Bob," she reported. "He seemed to think I was sympathizing with the animal too much. But it was always done to get the 'Fight' settled, and almost always worked."

One year they had a bear, bought from a woman at Gum Neck, North Carolina, who had hand-raised it. "By the time we got it, it must have weighed between twenty-five and forty pounds. . . . Bob decided to give the bear to Snookie as a pet. They were great buddies. Often at night, we could hear Snookie 'laughing' and the bear grunting, as they played into the wee hours of the morning." Ultimately the bear went mad and ran around the cage trying to bite Snookie and everyone else; then he died. Mae never thought of rabies at the time, but she has wondered since if . . . At any rate it was all right. Mae was not so lucky later, when Snookie threatened the two-year-old Velda Mae, and Velda's mother rushed in to remove her from danger. Snookie, who was tied up nearby and had somehow managed to reach the baby, grabbed Mae's leg and chewed on it, she said, "like a hungry neolithic man might have chewed on a dinosaur bone." The leg took a long time to heal, and gangrene set in. Mae refused for some weeks to speak to Snookie, and he suffered. He would approach her and whimper piteously, holding out his hand pleadingly, begging for forgiveness, but she would not forgive even when Bob took up the cause.

"Mae, he's so pitiful, speak to him!" said Bob.

"No, never again!" said Mae. "He didn't have to do what he did! I realize he's an animal, but he's not that stupid!" She turned to Snookie and said, "No! You're a bad boy! Look what you did to me!" and pointed to the bandage on her leg. Snookie came over and kissed the leg. . . . Of course, Mae gave in then, and after a while they were good friends again. He never again even tried to bite her, though he did bite Bob, more than once.

It was with an eye to breeding Snookie that the Noells bought a little female chimp they named Suzy Q, but the two animals never hit it off. Too late the Noells were told that it is better, when trying to breed chimps, to get a large female that can hold her own against the male. She should be older than her prospective mate, it was said. Later Suzy had a lot of children, but never by Snookie. In any case the next chimp the Noells procured, through the New York animal dealer Henry Trefflich, was the male Joe, who weighed nearly a hundred pounds and was trained as an entertainer. Joe became the most famous boxing—or wrestling—chimp in the Noells' outfit. Even apart from that, his training made him perfect as a show animal, for he rode a bicycle, a tricycle, and a scooter; he could jump rope; and he was perfectly at home in clothing, which included tennis shoes. Mae took him for exhibitions around schools, where it can be imagined how popular he was. Fascinated, she watched how cleverly he adapted his tricks. One, for example, was to cycle around and around the room until his trainer dropped his hat. Joe would then scoop up the hat on his next round. But one day Bob was showing him, and unfamiliar with the routine, he neglected to drop his hat. Joe went around and around, waiting for that hat. At last he dropped his own hat and picked it up when he came around again. The watching children applauded him thunderously, and Joe, without being prompted, simply incorporated that bit of business in the act from that time on. Mae knew that chimps value applause and will do a lot to get it, but this really impressed her. She was even more impressed when she saw that Joe, riding his bicycle around the Noells' campgrounds during a recreation period, introduced an improvement of his own into his customary game. He usually rode the bike down a hill nearby, jumping off before it crashed into a bank at the bottom. One day he didn't ride straight down the hill but followed the ground's contours so that he—and the bicycle—arrived together at the bottom of the gentler slope. How many human children, she asked herself, would have figured that out?

Mae Noell was changing. It was only natural that a woman who married at seventeen should change and develop as she grew older, but what happened to Mae was especially interesting, because what she turned into was a scientist. She watched the apes. She asked herself why they did the things they did. She studied their reactions just as she observed their eating and sleeping habits. One could hardly live in such close relations with the animals without learning a lot about them, but Mae went further than that. She became not only a student but a scholar: she read everything she could find on the subject—not that there was much at that time—and she did a lot of inventive thinking. As a result she wrote to Dr. Robert Yerkes of Yale, who was well known in her circles for his interest in and work with the great apes. Thus began one of the most remarkable correspondences ever preserved, even in Mae's well-stocked archives.

According to Mrs. Roberta Yerkes Blanshard, Yerkes's daughter, such correspondents were an important part of his work. "He couldn't very well have owned so many animals himself, especially gorillas," she explained, "but the knowledge these women had, even when they weren't official scientists themselves, was invaluable. So there he was, writing letters to and getting letters from all these women who owned apes, all over the country."

It was August 20, 1950, when Mae wrote her first letter to Yerkes. What inspired her was a visit she and Bob had made, more or less casually, to the Yerkes Laboratory of Primate Biology at Orange Park, Florida. The Ark was showing that week at Valdosta, Georgia, not far from the state border, and the Noells had come over to get new tires at Jacksonville. Orange Park was close by, and they had heard that the laboratory had a "huge" collection of chimps, so just on the chance, they telephoned and asked if they could drop in. They talked to Dr. Henry Nissen, and he cordially invited the family. It was Keith Hayes who took them around; he and his wife, Cathy, were partners in the enterprise of teaching a young female chimpanzee, Viki, to utter three words. Hayes noticed how much at home the ten-year-

77

old Velda Mae seemed among the chimps, and he suggested that the Noells might like to meet Viki and hear for themselves how she said "Mama," "Papa," and "Cup," in a hoarse whisper. The encounter was a great success; Velda Mae and Viki got on fine. Later the Hayeses came to see the Noell's Ark show, naturally bringing Viki along. Altogether it was a happy encounter, and Mae wrote to thank Dr. Yerkes for running such a fascinating ship. She told him that they now had a complete collection of great apes—six chimps, a gorilla, an orangutan, and a gibbon. (The gibbon is sometimes listed as a great ape, sometimes not.) The gorilla, said Mae, was a baby and, though a female, was named Goliath, a strange choice which probably owed itself to the fact that it is, as noted before, notoriously difficult to sex infant gorillas. Mae thought Goliath was probably a mountain gorilla, though, as she admitted, she couldn't really tell. The baby had webbed fingers; did that mean anything?

Yerkes at the time was compiling a census list of all the gorillas in captivity in the United States and Canada, so anything Mrs. Noell could tell him about the species was bound to interest him. Also, he was a kind man with a habit of instructing wherever instruction was called for. He replied at some length on the subjects she had raised; not only did he discuss the webbed fingers (no, he didn't think it was a characteristic of the gorilla, no matter which subspecies), but he was grateful that she had described Goliath's rash and slight illness. Mae thought it quite possible that it was chicken pox because it looked exactly like the chicken pox her two children had had earlier that same summer. Yes, said Dr. Yerkes, very likely that was what it was, and he was much interested in the possibility, its being the first time he had heard of such an illness in a gorilla. He was honestly interested in everything Mae wrote about her apes and was meticulous in his replies to her various questions. No, he did not think she was correct in supposing that there were only a thousand gorillas left in the world. Perhaps that number might apply to mountain gorillas, he said, though nobody could be certain, but if one counted the far more plentiful lowland gorillas . . . To his knowledge, he said, nobody else had ever used a gorilla in the way the Noells proposed, in a traveling exhibit. He wished Mae luck in her "unusual enterprise" and assured her that he would continue to be interested.

For her part, Mae collected all the Yerkes books she could find and read them with rapt attention. Not that she had a lot of time for reading. The

Noells had acquired another gorilla, a male that was larger than Mae had wanted, though Bob liked them big. This animal was named Ambam, or 'Mbom, depending on one's choice. Unfortunately, during the Noells' absence to pick up the animal, their orang fell ill and died, as did the gibbon—"though she was so disagreeable," wrote Mae, "it might be just as well." The orang's loss was deeply felt, however. In a happier vein, Mae wrote at length about Ambam's behavior and Goliath's games with the chimpanzee that had become her cage companion. Goliath imitated Ambam, she noticed, and hit herself in the chest as he did. The Noells had also acquired an infant mandrill, and when the weather was fine, all these young animals were let out to play on the grass. Goliath was adorable, said Mae; she just loved her to death. She now had twelve teeth, and Ambam had twenty. Much as Mae had always been fond of the chimps, she loved the gorillas even more; Goliath was so *darned* sweet.

Yerkes must have read all these letters, discursive, gossipy, confiding, beautifully written in Mae's copperplate, with close attention. For a period, when Mae was composing and illustrating and generally working herself to a frazzle over a little newspaper entitled the *Noell News*, much of which she printed by hand, he read all of it and admired it greatly. He appreciated the drawings of chimpanzee hands, feet, and facial expressions, with one recognizable portrait of Bob and his favorite wrestling ape, Joe. (Some of these illustrations were executed by their friend Jay Matternes.) When Mae felt forced to give up this publishing venture as too time-consuming, Yerkes wrote expressing his regret that it was closing down, and one can see that he meant every word. Sometimes he suggested books that would help her in her knowledge of primates, as for example *The Mentality of Apes*, by Wolfgang Köhler, and *Toto and I*, by A. Maria Hoyt. He repeatedly reverted to the subject of little Goliath's unsuitable name, pointing out that it was unfair to the animal herself, Mae's precious little girl, to continue under such a grotesque label. Undoubtedly the reason Mae held out so long against this objection was her conviction that apes got used to their original names, and to change them was bound to be confusing. At last, however, she submitted. She would have a name contest, she wrote, and would change Goliath's name to some more suitable one if Dr. Yerkes would consent to be the godfather. Had he any suggestions?

Readily he accepted the challenge. Yes, he would be glad to stand as Goliath's godfather, though he would like to know just what the spiritual duties would consist of . . . and he had ready several suggestions. What of Golly Girl or Gal? Fine, said Mae, and without more fuss Goliath was rechristened, Golly Gal, with—one supposes—little confusion on her part.

Mae read assiduously, not only primate books but anything that had to do with natural science, and often she asked Yerkes questions that obviously would not interest the people she usually encountered. The subject of food for the travelers on the *Kon-Tiki*, for example, captured her attention. This plankton—might there not be other foodstuffs in the universe that could be profitably given to captive creatures that didn't necessarily flourish on the human diet? She had often thought that her charges might do better on their own kind of regimen, if only one could be sure what it was.

"You seem to have an ingenious mind," Yerkes wrote, and it was true. Also, Mae was an excellent observer and was careful to tell him everything she could about her gorillas' behavior. After reading Yerkes's book entitled *The Mind of a Gorilla* (an animal he had studied in Florida), she confessed that until then she had thought of inviting him to see Golly Gal and Ambam for himself, but now, having read the book, she thought he had really observed just about everything she had to offer. Except, perhaps— had he ever heard chimps, gorillas, and orangs laugh out loud? With real guffaws? Because the Noell apes did laugh like that when the Noells romped and played with them. The little chimp Kongo preferred such romps even to eating, and Golly Gal loved to be tossed into the air and caught again like a human baby. As she went up, she would gasp gleefully, then chuckle when she came down. Ambam had made up for himself a game she described as blindman's buff, pretending he could not open his eyes and staggering around. If she had known, it was not an original game with Ambam; this writer has seen gibbons and human babies playing it.

An alarming accident had befallen a friend and colleague of the Noells. Eagleson, the friend, had a traveling show that included a family of rhesus macaques, the Indian monkeys often used as laboratory subjects. The male of this group, a large, powerful animal, had caught Eagleson in his cage and knocked him down. At that same moment the cage's steel door, descending, hit the man in the ankle, breaking it. Now it was feared that he would not be able to take his outfit out on the road that year. Spare Mae from rhesus, she said with an exclamation mark. Give her a chimp anytime. You can reason with a chimp. . . . In reply, early in 1951, Yerkes thanked her for her careful descriptions of gorilla behavior, especially of the games they made up, and of orangs laughing. As he said, his chances to make such observations for himself were limited. As if to prove his sincerity, he marked a passage in her next letter with red crayon. She was asking if he had ever known gorillas to *vibrate*. Both her gorillas vibrated, usually after a romp; it seemed to mean that they were feeling extreme pleasure. The Noells were acquainted with a Miss Hunter who looked after a gorilla in Ringling's circus, and Miss Hunter, too, had noted vibration in her animal. She said it could be likened to a cat's purring rapidly, only without the sound. At such times, wrote Mae, Ambam's tummy vibrated and he held his breath, letting go at last with a gasp. This happened only when he had been romping, and it indicated great pleasure; it was rather like a minor convulsion and could be called laughter.

In a sudden burst of confidence Mae told Dr. Yerkes, after cautioning him that she would not say this to just anyone, that she brought up her gorillas exactly as she had done her children, with plenty of love and kisses and cuddling, and they responded accordingly. After all, she *was* their mama, and that is how they obviously felt. She hoped this confession would not "repulse" Dr. Yerkes; needless to say, it did not. But shortly afterward, in April 1951, she wrote in distress that Golly Gal was ill.

An anxious nurse, she kept a record of the gorilla's temperature and general condition day by day in a list that she sent to Yerkes. On May 25 he replied, thanking her for her very interesting, if extremely discouraging, report on Golly Gal. Though it might sound cruel, he said, he thought that the little gorilla would prove a bad investment; her condition reminded him of the experience he had had with one of his first chimpanzees, the ill-fated Panzee. She was never well, and after she had died of tubercular infection, the autopsy showed her intestinal cavity to be "a tubercular hotbed." The report on Golly Gal sounded similar. He would be surprised if she came through this siege a healthy ape. He hoped the Noells had not bought her as a healthy specimen.

On June 8 Mae wrote that Golly Gal died six days after Yerkes's letter arrived. They were having the body autopsied. Before her death a baby specialist of Tampa had examined her stool and so on and expressed the opinion that the case was not hopeless, though the specimens proved full of tubercle bacilli. However . . . There would *never* be another Golly Gal, wrote Mae. Yes, they had paid five thousand dollars for her and had got her as a healthy specimen; they couldn't afford to lose that kind of money, but they had not insured her with Lloyd's because it was so expensive. Bob was still talking about getting another female as a mate for Ambam, but if they did, Mae said, she was determined not to get another baby; it hurt too much when one lost them. She was sending Golly Gal's X rays to Yerkes.

He wrote that another of his lay correspondents, Mrs. Gertrude Lintz, was at that moment struggling to save the life of a very young gorilla that had been brought to her by a ship's captain. Dr. Yerkes had hopes that she would succeed in bringing up this one, as she had had experience and her husband was a doctor who could help her. He added a few words of advice to the bereaved lady: The next time the Noells bought a gorilla—if such was their intention—they should try to get one at least three or four years old that had not been too long in captivity. In Africa, he said, the longer they were kept, the more likely they were to suffer from malnutrition.

Mae wrote that Ambam now proved to be suffering from hookworm, and so, not surprisingly, was Bob. Bob, in fact, had other parasites and had spent five days in the hospital, where the doctor told him that most of his trouble was "nerves." He had not taken a holiday, she said, for twenty years. The Noells might, therefore, travel around the country, looking at gorillas. Mae was sure she would love that, and it would make a nice change (one wonders how) for Bob. At the moment, however, she was preoccupied with persuading Ambam to take his worm medicine. Bob was less of a problem in that respect.

In September, however, Yerkes gave her news that interrupted their plans. Mr. Said of Columbus, Ohio—always a good address for primatologists, for it was there that Colo, the first gorilla ever born in captivity, was conceived and delivered—had returned from Africa, bringing five young gorillas with him. At least, so it was reported. Presumably these animals, or some of them, were for sale. In addition, Yerkes happened to know, Frank Buck of Camden, New Jersey (the Bring-'Em-Back-Alive Buck), had two or three young gorillas on his hands. Yerkes naturally heard these things, as he bought apes for his laboratory when he had the funds to do so. There was no reply, so he wrote again in January 1952, asking if the Noells had any new gorillas. He was at last getting out his census report, he explained, and he wanted to bring it up-to-date. At last, in February, he heard from Mae, thanking him for having sent the census report. She assured him that she felt honored that the Noells headed his list, but she had sad news: Ambam had died in November. They did have a new gorilla, however, a male bought from Bill Said. He was named M'Jingo.

People who have owned and loved dogs usually swear, when a dog dies, that they will never have another. But they almost always do get another, and the Noells were the same with their gorillas, with perhaps more inducement than mere sentiment to replace their pets. Gorillas do help round out a circus. To be sure, they could have carried on with mere chimpanzees, but the occasional gorilla added spice. This writer saw the Noells' show in the 1970s and can attest that Tommy, the three-hundred-pound (or was it six hundred pounds?) male gorilla, was the most impressive member in the troupe, even though he did no tricks and just existed, as did the red-haired orangutan family that lived with the Noells at the same time. Bob really loved his gorillas and his chimps as well. He never held a grudge against the chimps even after his notorious accident. It occurred in March 1953 in Heidelberg, Louisiana, while he was settling his performers into their temporary home, preparing for the show, and Joe, the veteran boxer, turned against him. Joe bit two of Bob's fingers clean off his right hand and badly mangled the left hand, though the doctor managed to save that one. Later Joe actually handed the severed fingers out of the cage to Bob's young son Bobby.

Bob never did, as they say, hold it against Joe. After he recovered, life at Noells' Ark went on much as before, Joe included. Indeed, during his anguished first hours in the hospital Bob got word out with certain directions. "I hastened to the lot," Mae remembered in her book, "to tell Bobby that his father wanted him to show the crowd his torn and bloody clothes, and to explain that they are trained wild animals and to let the people walk past Joe's cage and look at him."

Once a showman, always a showman—at least until he retires, and Bob was not to do that for many years more. Officially the retirement took place in 1978 and 1979. The Noells went to live permanently on the plot of land they had already prepared in Florida; it was named the Chimp Farm. And the apes retired with them, there to die in their own good time.

6. PENNY

THE ROAD TO THE GORILLA FOUNDATION outside San Francisco is hilly and wooded, and the foundation itself is well hidden behind a gate that doesn't bear the right name.

"Because we don't want to be overrun by curiosity seekers," explains Francine ("Penny") Patterson, the foundation's moving spirit. She occupies a house on the side of a hill very close to adapted trailers that hold Koko, the elder gorilla, and the younger Michael. Around houses and trailers stretches a large apple orchard that has been allowed to go its own way, and on the autumn day when I penetrated the leafy hideaway, apples lay everywhere on the ground, and a cidery smell filled the air. A black-and-white Newfoundland dog ran out of the house, barking, to be followed by a Manx cat and, at a distance, an inquisitive kitten. Penny Patterson, a striking young woman with long, flowing blonde hair, wearing blue jeans and a shirt, greeted me near steps that led up to a screened veranda, the screens being of strong chain link fencing, at one side of the front trailer.

Through the agency of the National Geographic Society (one wonders where primatology would be without the help of that body), the Gorilla Foundation got its first good PR push in October 1978, when the magazine *National Geographic* featured an article complete with photographs, entitled "Conversations with a Gorilla." In this long piece Ms. Patterson rapidly outlined the events that led to the household on the hill and all the concomitant details of life with learning anthropoid apes. A more detailed account is to be found in the book *The Education of Koko*, by Francine G. Patterson and Eugene Linden, and there is more in Penny's short biography or curriculum vitae. She was born in Chicago in 1947. Her father was a professor of educational psychology at the University of Illinois—he is now professor emeritus—and has written several books on counseling and psychotherapy. It was a large family, with three girls and four boys. Francine especially was devoted to animals—"Oh, yes, forever!" she said when I asked her. It was natural, therefore, that when she elected to major in psychology, she should lean toward animal studies. She took her A.B. in psychology at Illinois, emerging as a member of Phi Beta Kappa and Phi Kappa Phi, the psychology sorority.

After graduation, she and her constant companion Dr. Ronald H. Cohn, a molecular biologist, drove out to California, where she enrolled as a graduate student at Stanford University and where something very important happened: She attended a lecture given by the Gardners of Nevada State. Dr. Allen and Dr. Beatrix Gardner had been working for five years on the new, exciting idea of teaching a young chimpanzee the American Sign Language, or, as it is called for short, Ameslan (or ASL). This particular language was chosen because it does not entail spelling out the words but uses gestures for them—beckoning for "come here" and holding out the empty hand for "gimme." Penny, like many other scholars, was fascinated. Not everyone has been convinced; but the skeptics' protests have grown fainter as the years move on, and Project Washoe—named for the young chimp—has moved on with them, from triumph to triumph, even though Washoe herself has disgraced the name of chimpanzee by mutilating with her teeth the fingers of a famous neuropsychologist. Before her days of notoriety as a finger biter, the chimpanzee was adopted for a time by Dr. William Lemmon, then a professor of psychology at Oklahoma State in Norman, but now she is in the custody of Dr. Roger Fouts, in Washington. Back in 1971 not many people knew much about the subject. After hearing the Gardners lecture, Penny, afire with zeal, read whatever literature she could find on it and soon hit on the idea of teaching Ameslan to another species of great ape, the gorilla. In fact, she chose for her doctoral dissertation the subject she called Project Gorilla. She took a course in Ameslan and began looking about for the necessary ape.

At that time it was not a quite impossible quest. The government had not yet passed its law decreeing that no wild animal can be imported unless it has been born in captivity and the importer can prove it. Today a gorilla of this description, if it could be found—a big if—would cost a purchaser seventy-five thousand dollars at least; lately a buyer offering a hundred thousand could not find one. Penny, however, did not face such odds, and her thoughts, naturally, first turned toward the local zoo in San Francisco.

Her adviser was Dr. Karl Pribram, that unfortunate neuropsychologist who was to become Washoe's victim; he, too, was interested in a gorilla project, though not the same one. He thought it might be possible to teach gorillas to use computers to express themselves (an idea that at last came to fruition, in a way, under the management of the primate psychologist Dr. Duane Rumbaugh of Atlanta, though he used a chimpanzee, Lana). Pribram wanted to see the gorilla colony in the San Francisco Zoo and discuss computers (used to test IQs in apes) and so forth with the zoo director, Ronald Reuther. He suggested that Penny accompany him on the exploratory visit, and of course, she did.

It was September 1971, and the gorillas' youngest infant, Hanabi-Ko, was on view with her mother, Jacqueline. Hanabi-Ko was born on the Fourth of July, and her name is Japanese for Fireworks Child, but it was soon to be shortened to Koko. At the age of two months she was making hard work of hanging on to her mother. She should have been able to ride on Jackie's back, and Jackie kept putting her there; but every time she slid off. Perhaps because he suspected she was weak, or perhaps on general principles, Reuther did not react favorably to Penny's request that she be permitted to teach sign language to the infant. He reminded her that gorillas are rare, endangered animals. Penny had to go back to Stanford without satisfaction.

At the time she was working on another primate, the gibbon, and its capacity for self-recognition. It had been demonstrated that chimpanzees *know* that they are chimpanzees; when they look into mirrors they recognize themselves. How do we know? Because if a sedated chimpanzee is marked on its forehead, when it wakes up and sees its image, it touches the actual mark on the forehead, not the marked image in the mirror. Would a gibbon do the same? No, as it turns out, but Penny was rather halfhearted in her work; she was too absorbed in reports given her by a friendly keeper at the zoo on the progress of Koko. The little gorilla, he said, was not doing satisfactorily on her mother's milk, and soon the authorities had to admit that she was definitely ailing. In December they "pulled" her—took her out of the cage and her mother's care—and hospitalized her for a while, after which she was moved into the Reuthers' house for special tending. Deemed out of danger after two weeks, but not yet ready for exhibition, she was moved into another human household, that of the Children's Zoo manager and his wife, for convalescence. (It was there, by the way, that I had my first glimpse of Koko, lying in a basket and swaddled in a baby blanket. As I

Belle Benchley, San Diego Zoo director.

Quinta Palatino.

Robert Mearns Yerkes, with Chim and Panzee.

Madam Rosalía Abreu.

Andrés (with orangutan Cachesita), the keeper at Quinta Palatino, who described the birth of Anumá.

Primate living quarters at Quinta Palatino.

Oil painting of Cucusa and baby Anumá, commissioned by Madam Abreu.

Anumá, the first chimp born in captivi aged nine and a quarter years.

Augusta Maria Hoyt and Toto.

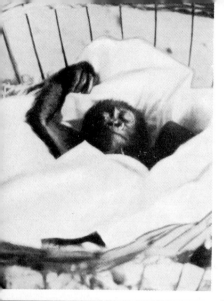

Toto: (counterclockwise from left) as an infant in Africa; taking her first steps; riding a tricycle; and with Tomás, her caretaker and constant companion.

Toto.

Gertrude Lintz with (left to right) Suzabella, Skippy I, and Captain Jiggs. (Insets) Captain Jiggs (top) and Suzabella (bottom).

Buddy, aka the ferocious Gargantua . . .

. . . at play with some of his favorite toys.

(Clockwise from bottom) The Noell's Gorilla Show, the purple ribbon awarded to those who wrestled chimps in the ring; Bob Noell with Kongo.

Mae Noell with Tommy the gorilla.

Bob Noell with Tomm

Stella Brewer, teaching Bobo how to use a nest.

William, Stella's problem chimp.

Raphaella Savinelli, who lived for a time
in Stella Brewer's colony, with Bobo.

Barbara and Tom Harrisson.

Jo Fritz.

Barbara Harrisson.

Barbara Harrisson.

(Left to right) Nadia Kohts, Sergius Tolstoy (grandson of Leo), and Robert Yerkes in Russia.

Nadia Kohts with her chimpanzee Jony.

remember, she was brandishing a rattle. Infant gorilla faces are mostly nose, and that is how I recall her.) Soon Penny's friendly spy told her that Koko was on view in the Children's Zoo nursery and that he thought it might be a tactful time for Penny to approach Reuther again with her request. She did. The keeper had been right, and Penny got her permission to teach Koko.

At first it was not an ambitious setup. Ron Cohn drove Penny to the zoo, and she was solemnly presented to Koko in the nursery, where the baby lay in the arms of her caretaker. She was a twenty-pound year-old infant, black save for the white patch on her rump typical of infant gorillas. She was put down on the floor and Penny made her first Ameslan sign, "Hello." Koko could not be expected to understand, but she patted her own head and Penny's and pulled her hair. Alone with Penny, she pretended to ignore her, but every time the girl moved she would rush over and try to climb her, as if afraid she would go. The interview ended when Debbie, the nursery girl, came back with Ron, so Ron could take photographs. While Debbie and Penny talked, Koko bit her new friend a couple of times. Taking this as a hint, the visitors decided it was time to leave, but Penny was back in the morning with a new toy, a wading pool.

Days passed. The zoo docents—volunteer guides—amiably accepted Penny's suggestion that they keep making simple signs to Koko, such as "drink" when she was given her bottle and "up" when they were about to lift her. At first she preferred men to women, ignoring Penny if any human male was around, never trying to bite them as she did Penny, but after a couple of weeks this tendency changed. She did not bite Penny so often and actually preferred being carried about by her new teacher. A sure sign of this transference was that whenever she was startled or surprised, it was to Penny's arms that she fled. Early in the morning Penny would take her for walks around the zoo.

"From the beginning of the project I had a dual role," she wrote. "I was a scientist attempting to teach a gorilla a human sign language, but I was also a mother to a one-year-old infant with all an infant's needs and fears. My initial problem was to establish rapport with Koko. . . ."

On these early-morning walks the little gorilla was afraid of all the big animals except the llamas, which she could easily intimidate. She was especially afraid of Bwana, the dominant gorilla, which reacted unfavorably to her. Perhaps this fear was healthy and rational.

Penny attempted to concentrate Koko's attention on her hands because they were, of course, to be of surpassing importance in the program. The teacher made up little games that called for using hands. One of them was to breathe on the glass of windowpanes until she had a fog, then draw little pictures—stars and other easy patterns—in the mist as long as it lasted. This game was a favorite of Koko's, and she tried to draw in imitation, producing scribbles and squiggles. The pastime was to have an interesting effect three months later, but for the moment it stopped there.

With Washoe in the Gardners' care, the people entrusted with her training were enjoined not to talk when they conversed with the chimpanzee in sign language, as this might confuse her and spoil the effect. Apes may be dumb as far as speech is concerned, but they are not deaf. However, Penny found it impossible to keep speech out of her program, since teacher and pupil were in constant view of the public, and the public naturally had a lot to say, loudly, when they saw an infant gorilla playing with a mysterious blonde. Penny resigned herself to the facts and simply made it a part of the training to speak the words as she signed them. There were other difficulties, too, in this goldfish-bowl school: Children would tap on the glass and tempt the pupil away from her lessons by showing her food or toys, though Koko could not reach them. Frustrated, she would close her eyes and spin around like a top, something that other young primates, including humans, often do. Big gorillas sometimes spin, too.

How long could Penny keep up her experiment? She did not know, but hoped for an allowance from Ron Reuther of four years, that being the amount of time the Gardners spent on training Washoe. Reuther's ideas were not so generous. He thought Koko should go back to her troop after six months, but the keeper who had helped nurse her held that for health reasons she should not go back for three years. In the meantime, Penny, hoping for the best, enlisted the help of two women, one deaf and therefore adept at Ameslan, and a docent named Barbara Hiller (who is still with the project).

At this point the university made Penny an allowance, salary money for another assistant whose "first" language was Ameslan, which was a relief. Nevertheless, they were parlous times. Controversy on the project swirled about the heads of both primates concerned. It was all the more lively because Penny was working not with a chimp—people were getting used to that idea—but with a gorilla, there being a lot of prejudice still against gorillas as ferocious beasts. People often asked Penny how she dared enter a cage with such a creature, let alone get close enough to mold her hands, as the process of shaping is called. Molding is important in the early stages of learning Ameslan, to give the child (or ape) an idea of what to do to express a concept.

It took Koko only a week to learn the signs for "food" and "drink." At least so the volunteers reported to Penny, who did not let herself believe it at first as she had not herself seen it. But they were probably correct. For the next few weeks Koko seemed to be making signs, but they weren't definite enough for Penny to accept them; she was on her guard against being too eager. On August 7 the teachers began a more formal kind of instruction, watching Koko carefully as they made signs, over and over, in the proper context, for "food," "drink," and "more." When it was time for Koko's bottle, before handing it over, they made the "food" sign, and if she did not respond, they signed, "What's this?" If she still did not reply, they folded her hand into the proper sign before handing her the bottle. It took her only two days to get the idea, and when she had it, there could be no doubt: She made the sign clearly and correctly. In fact, she sometimes made it without being asked or shown any drink, using the gesture to demand more goodies. That was an exciting day for everyone concerned. Koko's interest flagged only toward the end when she had eaten all she possibly could; her pleasure in communicating grew visibly less toward bed-time. Next day she proved that she knew what the sign meant by using it several times as she watched workers cleaning her cage and removing bits of leftover food. After the breakthrough she learned many more signs very quickly.

"Barely into the second month of training she moved from one-word expressions to two-word combinations—somewhat more quickly than Washoe had," said the proud teacher. Throughout, the scientist in Penny Patterson has warred with the gorilla chauvinist, and the scientist does not always win out.

"In all, during the first two months, Koko used about sixteen combinations," she said.

Some critics of Washoe complained that she didn't ask questions, but Lucy, the brilliant young chimp in Norman, did. She would inscribe in the air a large question mark for "What is that?" and point to some unknown object, as did most of her fellow students in the chimpanzee colony there. But Koko used a different mode of inquiry. I quote Penny: "From the outset

Koko spontaneously used eye contact and gestural intonation to phrase questions, a form that is considered legitimate in Ameslan." Penny first noticed this in the third month, September, after the true signing had begun. She was blowing mist on the windowpane in the old game, urging the gorilla to draw while it lasted, and they both did so as the mist disappeared. When it was gone, Koko pointed to Penny's mouth and touched it with the index finger while looking into her eyes. Taking this as a request for more window mist, Penny blew at the glass as usual. This time Koko tried to imitate her, getting close to the window and licking the glass. She did manage, probably by simply breathing as she did so, to produce a little fog, and she drew on it. Later that day, trying to make mist, she even imitated the puffing noise Penny made while blowing. A week later she elaborated on her new method of communication, pointing first to the glass, then to Penny's mouth, then to her own mouth, and then to the glass again. It must have given a strange, if triumphant, feeling to Penny.

In the meantime, as Koko grew older, she got more skillful. At fifteen months she could turn on a faucet and get drinking water for herself. There was, inevitably, difficulty in using her hands while signing because a gorilla hand is not quite like a human's. Koko's thumb was much shorter than ours and farther down on the wrist, so gestures had to be altered to meet this difficulty, and allowances made.

Penny and her assistants kept careful record of the words and phrases used by Koko. For the first year and a half her record of learning pretty well matched that of Washoe, against whom Penny measured her constantly; she produced one new sign learned each month. At age one and a half Koko's conversations could hardly be dignified by that word—rudimentary Penny called them. "Up," the gorilla would demand when she wanted to be lifted, and "food nut more," which hardly needs translation. But a year later showed definite improvement. Penny recorded a conversation that took place while she was playing with the gorilla, spinning her around on a tabletop.

"Tickle," signed Koko. Penny tickled her, or rather made the sign for "tickle" on her hand, and asked, both with signs and voice, "What do you want?"

Koko: "Out key."

Penny: "What?"

Koko turned and looked out the window. Penny got out her keys.

Koko: "Open sweater key."

They keep Koko's sweater locked in a cupboard. Penny held up the key to the cupboard.

Koko: "Key."

Penny gave her the keys.

Koko: "Key, key."

She shook them up and down.

A year later Koko would sign, "The key, the key," when she wanted Penny to open a door, or window, or cupboard, and then make the sign "open."

She saw their adopted cat (known as Koko's cat) outside the window, and Penny called, as one does, in a high-pitched voice, "Here, kitty, kitty, kitty." Koko, who had not heard this phrase before, looked at Penny in surprise and later signed, "Cat, cat, cat, cat." Then she signed to Penny, "More there," and took the girl's chin in her hand, pointing to her own chin and signing again, "More more there."

"More cat say?" signed Penny.

"Cat," repeated Koko.

Penny called again, "Here, kitty, kitty, kitty," and Koko seemed pleased and satisfied.

By this time a change had come into the life of teacher and pupil. Penny had never got used to working with Koko in the public eye, as she was forced to in the Children's Zoo, and she began to think that a trailer would be better. Ron Reuther agreed, on condition that they find a trailer that fitted in next to the gorilla grotto. He and Penny went out looking for what they needed. Cost was also an item, naturally, and they were pleased to find a rather shabby, secondhand trailer the right size, partly furnished, for two thousand dollars, which they agreed was a bargain. It was installed next to the office trailer, and Penny was just trying to accustom Koko to it when circumstances demanded a complete move. In June 1973 Koko knocked too hard on a window in the nursery and broke the glass. Then Reuther insisted that the little household move outright to the new home. This turned out to be not only a relief but the reason for more development. In her trailer Koko was more in touch with the world outside; for example, a friendly policeman amused her by making clicking noises, and she amazed Penny by imitating him.

"Since gorillas are not supposed to be able to imitate sounds at all, I was reluctant to believe my ears," she wrote. "Subsequently, though, she has imitated other unvoiced noises."

But this comparatively halcyon existence was disturbed by the threat that some authorities in the zoo wanted Koko reinstated with the other gorillas. Their reasons were that as things were going, Koko would soon be too humanized to be willing to mate, and after all, the breeding of rare and endangered species was the zoo's main preoccupation. Besides, they argued, the older Koko got, the more dangerous she would be to work with. They didn't want to be responsible, they said, for a strangled Penny Patterson. Of course, Penny didn't agree with their reasoning. She appealed to the benefactor Carroll Soo Hoo, the man who has donated many gorillas to the San Francisco Zoo. He often went into their cages to play with them, no matter how large they grew. So much, she thought, for the safety argument, and he agreed with her. As for the other argument, she was sure that if they could find a young male gorilla when the time came, Koko would be quite willing to mate with him. *When* the time came; just then Koko was still only three years old.

"If she has to go into a cage," vowed Penny, "I'll go in with her. They simply can't do this to Koko."

By now, as she admits in one of the most interesting passages of her book, her attitude had changed. She was no longer coldly objective, if indeed, she ever had been. Koko had begun to "get to her," dependent and affectionate and engaging as any human baby. She wrote:

> Caring for her entailed most of the joys and stresses of parenthood, and like a parent, I was endlessly fascinated by her development and charm. She cooperated with chores, assisting in cleaning and handing me items on request. She imitated my every move, from talking on the phone (Koko even opened and closed her mouth and huffed and screwed up her face) to grooming her fingernails when I did mine . . . she'd wrestle with and kiss her dolls with loud smacks, tickle my ears or make me a part of her bedtime nest by arranging my arms around her, gently pushing my head into place, and lying down and cuddling. . . . As her vocabulary grew and Koko began to use words that revealed her personality, I began to recognize sensitiveness, strategies, humor, the stubbornness with which I could identify. It was the realization that I was dealing with an intelligent and sensitive intellectual that sealed my commitment to Koko's future. My knowledge of Koko's vulnerabilities made the prospect of returning her to the gorilla grotto unimaginable. By the time Koko was three, I was afraid that if that happened the trauma of separation would kill her.

Also, she was proud of Koko. She wanted to see how far the experiment could be carried. The end of it became an intolerable prospect. She was not alone in this feeling. While some of the authorities thought that no good could come of her continued involvement with Koko, others thought it would be a pity to bring it all to an end. For a while, however, the opponents had their way.

Reuther left the zoo, and an interim director decreed that Koko should be put on exhibition at least part of the time, say, a couple of hours a day. Penny promptly announced that she would go into the cage, too, and she did so every day along with an assistant. It was a disagreeable experience because the cage was noisy and some people threw things in. Besides, it was drafty, and Penny worried about Koko's catching pneumonia. She put a sweater on the animal, and some of the officials promptly objected, saying that animals in the wild do not wear sweaters. In vain Penny argued that animals in the wild don't live behind bars either.

On this unsatisfactory basis things proceeded until one of Penny's champions, the zoologist Paul Maxwell, made a suggestion. Kong, a young male gorilla belonging to the Marine World in Redwood City, near Palo Alto, was not much older than Koko and would make a good companion for her, long enough, at any rate, to reintroduce her to her own world. Nobody had plans for the animals to mate—it was far too soon—but it seemed a good idea, and at least it put an end to cage duty. However, Redwood City is down the peninsula from San Francisco, and one gorilla had to travel the whole distance to meet with the other. The zoo didn't want to risk Koko's life on the road too often, while the Redwood City people feared that Kong might catch some disease if he commuted to the city to see his friend.

The anti-Penny authorities were pushing for a solution, saying that something had to be done soon; more and more they feared that Koko was becoming hopelessly humanized. They talked of lending her out to some other zoo and asked if Marine World would be willing to buy her. Penny decided that neither of these things must be allowed to happen. Why shouldn't she move Koko to the Stanford campus, trailer and all? There they would be within easy reach of Redwood City, Koko would be away from that threatening cage, and everybody would be happier. And it was done. Pribram helped to buy the trailer from the zoo. Dr. Richard Atkinson, representing Koko, Penny, and Ron Cohn, managed to get a two-year grant (for speech therapy) from the Spencer Foundation, and the move was made. Koko did not enjoy it and kept signing all the way along the road, "Go home." The first night on the campus, frightened by the strange noises, she kept waking and crying, and Penny stayed with her every night for a month, until she got used to things. After all that Kong did not come over very often. He was getting too big for his attendants to handle. Indeed, for a time Marine World tried to sell him to Penny, but she and Ron decided against it for the same reason Marine World wanted to get rid of him.

Even so, things were better at Stanford. Away from the zoo, Penny could think more clearly of the future. Carroll Soo Hoo, as a self-appointed go-between, asked Ron if the couple would be willing to buy Koko if possible. Certainly they would, said Ron, but it was not to be so simple, even if finding the enormous sum required could be called simple. San Francisco now had a permanent zoo director, Saul Kitchener, who had come from the Lincoln Park Zoo in Chicago and had always had a special interest in gorillas. On the subject of Koko he was adamant. She could go to Penny and Ron, he declared, only if they replaced her with another female gorilla. Though this may seem a fair and equitable arrangement, they all knew it was not so easy to do; the ban on the importation of wild animals had taken its toll on the gorilla supply in the United States. Ron and Penny found an available female at the Yerkes Primate Center in Atlanta, but she was arthritic, and Kitchener turned her down. Another could be got from Honolulu, but Kitchener said she was too old to breed. At last they heard of two gorilla babies for sale in Vienna for twenty-eight thousand dollars. Vienna claimed they were captive-bred, and though Penny and Ron were sure this was a lie, they were willing to accept the fiction. The little female could go to San Francisco, they decided, and they would keep the male. So, though they had only seven thousand dollars in hand, they said they would buy the animals, which arrived September 9, 1976.

Unfortunately the female died within a month, of pneumonia. The male, named Michael, moved in with Penny and Koko, bringing with him that hideous debt still outstanding. However, during her struggles Penny had learned a lot about the power of the press, and now she used it.

"On March 9, 1977," she wrote, "the *San Francisco Examiner* published an article about the uncertainties of Koko's future. The article reflected the sense of urgency I felt about Koko's future, and quoted me accurately as saying that I felt Koko would die if she was returned to the gorilla grotto." To be sure, it also quoted Saul Kitchener as saying that he had never heard of anything like that happening, but as Penny pointed out, there were no precedents involving language-using gorillas on which either side could base its feelings. It is a fact, however, that apes *have* died after abrupt separation from their caretakers. In any case the article touched hearts among the public; donations amounting to three thousand dollars came pouring in and went straight into the gorilla fund. There followed a Save Koko campaign with bumper stickers and all that, which also helped the war chest. In time, Ron, Penny, and the lawyer Edward Fitzsimmons created the Gorilla Foundation, a nonprofit organization that was to hold trust over Koko, protect her interests, and abet the study and preservation of all gorillas.

All this generated considerable relief, but Penny was run-down from nursing that dying baby gorilla and general worry, and she fell ill. It was feared that she had Hodgkins Disease. Fortunately it was a false alarm, but Kitchener was not; he still hovered on the horizon like a threat, until the mayor of San Francisco lost patience and declared that the zoo must allow Penny to buy Koko and be done with it. Kitchener gave in, though unwillingly, and Koko, at last, belonged to the foundation.

It was summer 1977, Koko was six and Michael four, at which point the *National Geographic* got interested and in its usual benevolent way stepped in to investigate and offer financial help. The *Geographic* had done something similar for the Gardners and their Washoe, and, of course, it had helped to finance Jane Goodall and her chimps. There was Dian Fossey and the mountain gorillas, as well as Biruté Galdikas and orangutans: one might say the *Geographic* cornered the market on ape ladies. It did an issue on Penny and the gorillas, with Koko resplendent on the cover. It all helped.

It was in 1984 that I fixed up to go out to Woodside and meet Penny, Ron, Koko, and Michael. We peered into Koko's quarters first, and she peered out at me, the stranger, and made signs interpreted for me by Penny, who replied to them as fast as they came. Koko sat behind the chain links that divided us with the dignity that goes with weight (she was about 250 pounds), but there was nothing else dignified about her. She was eager. She opened her mouth wide, fixed her dark eyes on my face, and stuck one finger behind her upper teeth in an urgent way. Penny explained: "She wants to see if you have any gold teeth." I, too, opened my mouth as wide as I could, and Koko scanned the exposed area with rapt attention. So did Penny, who reported to Koko, fingers flying, "Yes, she has several gold teeth. . . . But what she really wants," she added to me, "is for you to take your teeth out. One of our visitors once did that, and she lives in hopes somebody will do it again."

I was sorry not to oblige, but I was able to answer Koko's urgent question as to what I had in my bag. ("She calls it a flower bag," said Penny, puzzled. "I think somebody brought her flowers once in a bag.") I didn't have flowers, but I did bring out a hairbrush, at sight of which Koko immediately asked for her comb. There was also a roll of Tums.

"Candy," said Koko. "Gimme candy."

"No, that's not your problem," said Penny. I said I had something else in my tote bag, a roll of Gelusil which I didn't want because it was constipating.

"That would be good for her," a new voice said from behind us; it was Ron. "She could do with something like that. Do you mind letting me have it? I'll give it to her." He had an opening in the chain link fence at the corner of the veranda.

Koko went over there to collect her candy, while Penny watched with a certain anxiety, and said, "I think she has an earache, but she hasn't said so. She keeps putting her hand up there."

"Eardrops?" I suggested.

"I suppose so," she said reluctantly. "We've done it before, but it's always a struggle. Ron, do you think it's warm enough for her to go out into the yard?"

If she wore her jacket and stayed in the sun, said Ron.

"What on earth is she doing?" I interrupted. Koko was advancing menacingly toward me, waving something long and thin.

Penny said, "Oh, that's her crocodile. Koko's afraid of crocodiles, I don't know why, because she's never seen a real one. She threatens us with that, too, all the time. It's a game."

Koko brandished her crocodile, and I recoiled as if in terror. She loved that game and repeated it several times.

"Funny, isn't she?" said Penny. "She's got a new terrorizer as well; one of the girls has brought her a dinosaur."

Koko went back to watch out of the chain links and viewed with some disfavor the antics of a kitten that had followed us up on the stoop and was now wrestling with my handbag. It was Koko's kitten, said Penny, and she seemed to think of it as a nuisance sometimes. She was now in a better light, and I exclaimed at the smallness of her feet. Gorilla feet, like their hands, are noticeably unlike ours, being stubby, with the big toe not aligned next to its neighbors, but down to the side, more convenient for gripping. Koko's feet were almost petite.

"I know," said Penny. "We take the same size in shoes."

Ron was inside the room now, helping Koko into her red jacket—a present from one of the volunteers. Since it had to be large to accommodate Koko's chest and arms, it was oddly shaped, but it did fit her. She played with one side of the zipper track, running it up and down without closing it. She went out into the play yard, which was fenced and filled with hanging tires and other playground equipment, while we moved around and kept her in view.

Penny told her to come over and sit on a stool in the sun. Koko looked about at the trees, the apples on the ground and the dog. Penny said, "This climate's not really the best for the gorillas. We're going to move if we can manage it. It isn't only the damp. This place is built almost directly on the San Andreas fault, and you can't help thinking about quakes. If we had a bad one, what would happen to the gorillas? We've thought of other places; the most appealing is on Maui in Hawaii—do you know that island? A lovely climate, and the authorities are friendly; we've been offered a lot of acreage. But we'd have to put in electricity and water, and it would cost *thousands*. We're trying to figure a way, but so far it's just a dream."

Koko kept her eyes fixed on us. Now she broke out into finger conversation. "She wants to play a game with you," said Penny. "She wants you to run."

"Me? I can't, Koko," I said. "I'm too old."

Inflexible, Koko repeated her demand. Feeling foolish, I obeyed by running up and down in front of the cage, while Koko watched approvingly.

"There now, wasn't that nice, Koko?" said Penny.

Koko ignored her. Taking off her jacket, she moved into the shade.

"No, get back into the sun," said Penny. "Oh, very well, you'll have to go back into the house. Ron?"

As Koko was removed, Michael was brought in for his turn in the yard, which he shared. I exclaimed at his beauty. At eleven he had not yet developed the heavy-looking muscle that characterizes full-size male gorillas, but his hair was beginning to show the white ends that would make him what is called a silverback. Even so he looked beautifully muscular, with a face like an Egyptian mask carved in obsidian. He stood as if posing, staring into the trees where the dog moved around among fallen apples.

"They call him the Tom Selleck of gorillas," said Penny, smiling. We went over to the foundation office in Penny's house, and she and Ron told me more about the male gorilla. He was even better than Koko at signing, they said, and expressed himself more readily to his teachers. But what interested me more was the information that Koko is learning to read. Penny explained: The only way in which Koko exceeds a human child in development is in her visual discrimination, which is excellent. For example, take the difference in a printed word such as "cat" and its neighbor "cot." Koko would spot that immediately, and such things are beginning to interest her.

"If there's an insignia on your dress or on a book, or a little thing like that, she can pick it out. She must have pretty good vision to do that. In fact, she'll notice that somebody's insignia is a bird and tell you so. She's really good at that game What's Wrong with This Picture. . . . She can find the little wrong things really well. I would think it's quite an adaptive talent, what with having to find the appropriate things to eat, that sort of thing, and maybe for social relations as well, because facial expression is really important for them. Watching her, even I can see how her expression varies between a dark, hostile facial configuration all the way to a happy expression, through all the gradations. And because gorilla faces are so dark, it might take more skill to judge their expressions. It would be very important to steer clear of an angry gorilla; you wouldn't want to tangle with him! They can pick up instantly on people, and this must be absolutely visual; it can't be smell because they are so often behind glass windows. They make judgments as to whether they're going to relate; it's got to be visual. Quicker than I could possibly do it; I can't judge a person in that amount of time. But they both do it, and most of the time they stick with it. Sometimes there are exceptions, but most of the time that's it. So the reading part is natural, I think. At first people will think no, it can't be, but why not when they're so quick at visual discrimination? Quite early in the reading thing when we were just playing around, Ron—he's really funny, likes to joke—well, he would take Koko's words and spell them wrong, make hard spelling errors like 'banama,' and she would catch it;

he'd ask her what was wrong with this word, and she'd find it. And I said maybe we'd better give her a little substitution test and see if she could put the correct letter in there, and she would do that, too. Some of her teachers have tried to get her to write a word, and she'd get two letters of 'frog,' for instance; they were kind of crooked but the *R* was pretty darned good. She'd even try to replicate the letters. She uses chalk. One day—I'd never done this before—I said, 'Can you do the letter *A?*' and she did it! I thought, Oh, my God! Where are we? Janet didn't know—she hadn't read my notes for that day—but the same day she was doing the alphabet with Koko and there was Koko copying her, distinctly copying words after her. So that's her ability. Then we got stuck with some letters, so we had to take things like *P* and *Q* and *T* and *F* and *M* and *N* that are real hard to distinguish. You can't, say, make a *T* versus an *N*, so we decided to take a word she likes, like 'tea,' "which stands for *T*, and 'nut' for *N*. She's taken 'Mike' for *M*, so she's made her little code, and you have to know her code. She changes it all the time. We thought *we'd* give her the code, but she's decided on a much better one. She liked 'peach' for *P*. We wanted 'potato,' but she likes 'peach' much better, so that's what she uses.

"Hearing is very important to her. She does a lot of her inventions based on the sound of words— 'obnoxious,' for instance, and 'You blew it.' There are no signs for these phrases in sign language, but we say them all the time, verbally. 'You blew it, Koko.'" Penny laughed. "So she's changed it into something incredible. When we say it, she jumps and waves her arms and blows so hard her hair flies—'You blew it.' And the 'obnoxious' reaction is a violent kind of knock." Penny knocked hard on her chair. "We didn't know what it was for a while. The part of 'obnoxious,' 'knock,' is the only part she can say. There's another sign we can't read, and we don't know what to do about it. The only thing I can think of is that it's the closest she can get to 'Don't know.' And it's possibly 'Don't know there,' meaning that she doesn't know what's there, hoping you will tell her rather than 'What's there?' Yet she used to say 'Don't know'; it came right off the top of her head then; we'll have to work it out, why she doesn't say it now.

"It's too bad to have only two animals. Of course, we hope for more, babies, if possible, and if we could move to a nice climate like Hawaii, we could perhaps get others on loan.

"But that's not all. The things you have to think of in a project like this! You have to be in a politically stable place, you have to have access to volunteers—schools mostly—a good climate. . . . There are so many restrictions I wonder if we're ever going to find the right spot. That place we were offered on Maui is just beautiful. It has a population of—what, Ron?"

"Three thousand."

They talked wistfully of the 680 acres in Hawaii. You could develop a wild animal park, said Ron. "But it's the water and the electricity."

For some time we discussed ways and means. PR all over the States? A motto in Honolulu: "Bring the gorillas to Hawaii!"? Pictures of Koko distributed to schoolchildren?

"It all came about in Hawaii because we had an exhibition there of the gorillas' art," Penny said. She brought out a number of photographs to show. "Oh, yes, they like to paint once in a while, Michael more than Koko, but they both do it. Here are some."

They were not, as I had expected, like paintings made by chimpanzees, with which I was familiar—but why should I have expected them to be like that? I shouldn't have been so racist. Koko's pictures were not like Michael's, for that matter. They use acrylic paint and have a free choice of color.

"But—why, they're representational, some of them," I said when I looked at Michael's efforts. Penny and Ron nodded, surprised by my surprise.

Penny said, "Here's a representation of the dog Apple that you saw." It was nearly a good portrait, nearly the conventional picture of the dog's face.

"And they *name* their pictures," said Ron proudly. "He named this one *Apple Chase*."

Penny added, "Koko names all our animals, really strange names, some of them. Here's a self-portrait of Michael. Here's a portrait of me. This is Koko's picture of a bird; if you look carefully, you can see a bluebird there. They did a whole lot of these for the art exhibit in San Francisco, and one of the critics said the only good things in the show were by the gorillas. Here are Koko's *Love* and *Hate*. She's not really all that keen on painting, but Michael loves it. I wish now I hadn't sold so many of their best ones, but of course, we needed the money. I asked Koko to do these two, *Love* and *Hate*. You see, *Hate* is dirty brown, but *Love* is pink and orange."

"Look at this one," said Ron.

"That's *Cat Trouble*," said Penny, laughing.

"Naming the pictures themselves," said Ron, "adds another dimension. The really remarkable thing about all this is the naming. A month or two ago we had an earthquake, a pretty big one, and next day we asked the gorillas to paint it. Michael used a very dark brown."

Penny said, "Here's one by Michael, and it's supposed to be blocks. I think he must be very disturbed about *Love*, because, look, it's all in blocks. Koko wasn't being terribly cooperative then."

"Do they use their fingers?"

"Oh, no," said Penny. "Brushes. Michael sometimes uses leather, balls, anything that will carry the color. He has always been good; he tried representational drawings, like the tree outside his window, when he was very young. Koko tends to do the drawing where it is—that is, if I give her a sheet of paper with some of my writing on it, she draws or paints over the writing. If I've made a list like a column, she draws a column, too. Here's one Michael did, trying to get a picture of the toy dinosaur." We talked a little about the dinosaur and Koko's trying to frighten people with it. "The thing is, she's such a silly." Penny laughed. "She does silly things, and you hear her laughing at herself. She's such a *character*."

I recollected the photographs in one of the magazines: Penny had put on a grinning mask, and Koko was grinning in imitation, pointing to her own exposed teeth.

"She thought I had gone crazy," said Penny fondly. "Gorillas have this funny chuckle, like this." She made a chuffing noise. "When she tickles my feet, Koko chuckles. Of course, I have to yell. She even chuckles sometimes when we crack a verbal joke. I think it grows from the contentment sound, you know that rumbling kind of purr? That grades into the laugh and then in the hysterical, uncontrollable laughter they do sometimes. When they're happy, they make that sound. She uses it sometimes to sing along. We have a music teacher who comes in, and Koko will make these vocalizations and go along with the teacher, humming and purring and sometimes matching the intonation. She does have a reasonable amount of control over her vocalization. Both of them have a funny sort of puff, a yes or no in their own language."

"Isn't there a kind of cough, too, in warning?" I asked.

"Oh, for sure," said Penny. "I call it a bark." She brought out some more photographs. "Here's Koko a day old, with her mother. Jackie died last year of valley fever. They look alike, don't they? Koko has a half brother at the zoo; you might go and look at him, and her father. She's got a full brother who looks very much like her; his name's Kubi. He would be about eleven now. He was devastated when his mother died. They didn't know whether he was going to live, he was so withdrawn—eyes sunken and hair all every which way. He took it hard; they all did. Gorillas mind their losses terribly, and she was the dominant female."

I said, "I wonder if Michael remembers his mother."

"Oh, yes," said Penny, "he does. We know because we asked him. He said he was with her in the woods, and some bad men came and hit her *here*, on the back of the neck. He saw the blood. After all, he was about two."

She began to put the photographs away, while I wrestled with the new concept. Gorillas as reporters? Historians? Well, why not?

Lately Penny and Ron have been stepping up their publicity about Hawaii, with good results. It is possible, really possible, that they and the gorillas can make it one of these days to that island in the Pacific.

7. STELLA BREWER: CONSERVATIONIST

AT FIRST GLANCE Ms. Brewer might apppear to be a copy of Jane Goodall. The comparison is obvious even to the women involved. Both work in Africa with chimpanzees, the younger Stella starting her career a few months after Jane had published *In the Shadow of Man*, a book that had a tremendous effect on Stella. Inevitably she wrote to the author, and the more experienced woman replied, helping the beginner with encouragement and advice. But there the resemblance ends. It is not simply a matter of difference between pioneer and follower, but one of background and education. Ms. Goodall, born in Bournemouth, started in her late teens as a newcomer to Africa, deliberately taking the chance offered by Louis Leakey, coolly making her observations in a scientific spirit remarkable in so young and inexperienced a woman, and learning the anthropological approach at the University of Cambridge even as she worked, snatching her degree, as it were, between sessions in the field. Stella Brewer was born near the wilderness of Mahé Island, the Seychelles, where as a toddler she was photographed riding a giant tortoise. You don't get much more exotic than that. She was six in 1958, when her parents moved with their three daughters to The Gambia, which is both a river and a country in West Africa, just south of Senegal. Her father, Edward, was and still is director of wildlife there. His domain, necessarily small, nevertheless includes a variegated fauna and flora for which he is responsible.

Gambia, the country, is a wriggly strip of territory that follows the river's contours. There seems no hard-and-fast rule against calling it simply Gambia, but Stella never omits the article "the" in her book *The Chimps of Mount Asserik* (or, in the British edition, *The Forest Dwellers*), and she ought to know. The Gambia is the preferred colonial name. In 1958 the country was still a British possession, just as Senegal was French. As long as the British were still in power, the forest with its many denizens had to be protected, and all the Brewers were caught up in the cause. Like most of their compatriots, they were fond of animals, not only dogs and cats but the less ordinary species. In the Seychelles the small Stella had adopted a fledgling bird. To be sure, the family cat immediately seized and ate it, but the child had tried. Her caring instinct was encouraged in The Gambia, where she and her little sister soon acquired pets. In the house were protected an African barn owl, a family of pouched rats, a porcupine—things like that.

Near The Gambia's principal city, Yundum, was a place of unspoiled woodland close to the airport, where birds and monkeys and antelopes and other fauna were to be found, and on Sunday Eddie Brewer would take his little girl there to watch the animals. It was an idyllic existence for a child, but no idyll goes on forever. Colonial British families followed a set pattern, and when the Brewers went on leave to England, they left behind at boarding school their two younger daughters. So Stella was kept out of paradise, except for short visits, until 1967, when her schooling was pronounced sufficient, and she returned to the old family routine, looking after animals in The Gambia. There were plenty of these around the house—a serval, a genet, hyena cubs, antelopes.

One day there arrived, in very bad condition, an infant chimpanzee. There are no native chimps in The Gambia. William had been brought all the way from Guinea in a box too small for him, tied around the waist to the box, cruelly chafed, terrified, and almost dead from the long journey by bumpy truck. The Ghanaian who brought him wanted a few shillings for his prize. Stella's father, to relieve the little ape from his misery, incautiously bought him, thus joining many others who are seduced into responsibilities of horrendous proportions. As Stella was to write, "We little realized what the consequences would be."

It took six weeks of careful feeding and cosseting before William, the chimp, bestirred himself and touched the Brewers' dog, which had mounted guard over him all that time. After this breakthrough, however, he grew and flourished, and inevitably became a genuine headache in the house. He investigated everything and stole, or broke, almost everything he touched. The Brewers loved him but were more and more distracted, until he managed—almost—to kill himself by drinking a bottle of turpentine. Something had to be done, and Eddie Brewer came up with a temporary solution. Two miles away from Yundum was a place known as the Abuko Catchment Area which, though he had never investigated it before now, proved to be a gem of natural beauty. All around the pumping station which was the reason for the area's existence lay a parklike tract of land thick with trees and vegetation, theoretically protected by a fence and a placard that said NO ADMITTANCE EXCEPT ON BUSINESS. A little river behind the pumping station poured its water into a lake. The place consisted of only 180 acres; but its vegetation was varied, and it seemed much larger. It also held a lot of animals. Stella's father wrote to the authorities suggesting that the area be turned into a reserve, and they agreed, provided Brewer would be responsible for it. Thus the Abuko Nature Reserve came into being. It held several sorts of monkey, and there were specimens of many other species of wildlife, including a leopard.

The nature reserve was created just in time to attract a new species, tourists to Africa, who came during the dry season in considerable numbers. With the help of his family Brewer gave lectures to these people, with slides and peeps at real animals, spreading the gospel of conservation (no fur coats, no exotic handbags, no lizardskin shoes), and collecting gate money to be used for the reserve upkeep. Naturally he also had William in mind, but he could hardly do what he wished, which was to build a dear little rustic cottage right there in the woods and leave the ape to shift for himself. It wouldn't do. Chimps, especially young ones, cannot live alone without dire psychological consequences. William had to live in a family group, and failing his own kind, the Brewers' group was elected. Edward Brewer must often have sighed with regret for those few shillings he had spent on William. But then, fortunately for everyone's sanity, Ann arrived. She was another young chimp from Guinea, and though Brewer was aware that it was wrong to pay for these animals and thus encourage Africans to catch more of them, he could hardly turn away this one. Afterward, however, the word went out that he would buy no more. Any chimps caught within the boundaries of The Gambia were simply confiscated by the government and—turned over to Brewer.

With Ann, William was a lot less trouble. In time they went to the nature reserve, where a fine big cage was put up, with plenty of swings and other toys. Even so, they wailed for human companionship. Anthropoid apes take a long time to grow up and become independent. Every day Stella came to the reserve and took them out of the cage to play around in the open. In so doing and without realizing it, she was developing into a genuine primatologist, observing and mentally noting the strangely human behavior of her two little charges. Then she took to writing down her notes. She reported how William, after watching laborers build the fences and enclosures that surrounded his and Ann's living quarters, was able to use a hammer, pliers, and even a spade.

The first confiscated chimpanzee, a small male named Albert, was sent to them by government officers and added to the couple already there; but Albert had not been captive for very long, and their enclosed life was not to his taste. As soon as he was introduced to Abuko, he escaped and ran away in the surrounding forest. Nobody saw him for a while after that, and he was given up for lost until, six months later, he turned up, alive and well and getting fat, but still uncatchable. He would hang around on the outskirts of the enclosure, watching the others. An older chimp, Cheetah, was sent to the Brewers; Cheetah, unlike Albert, stayed put.

Every day Stella continued to take them out for their airing, just for a change. They never ran away, never got lost, even though the sight of Albert, free as the air and watching from the trees, must have given them ideas. But the young chimps still depended on Stella at that period. As for her, she was becoming more and more an expert, trying to teach them what foods found in the wild might safely be eaten, for she was thinking of the future and their possible reintroduction to the wild. The trouble was that she herself didn't know what was safe to eat. They all would have to learn together, a risky business, young primates lacking any natural knowledge of poisons to avoid.

Then the troop acquired Tina, another confiscated ape. She proved to be a real prize because she was about six years old and had evidently had some experience in the wild. She had suffered cruelly and, as a result, was very shy. Stella at first had to leave food for her at the foot of whatever tree she was occupying, but she did know what fruits and berries to eat. Also, alone among the others, she made nests for herself to sleep in, a skill the smaller chimps had not acquired. In time, Stella hoped, the little ones would imitate Tina and thus be better prepared to shift for themselves. In the meantime, the girl herself, whether or not she realized it, was getting prepared for what was to be a large part of her life. Baby chimpanzees can be very insidious.

Happy, a very young ape, was the next acquisition, and Tina immediately adopted him. Yet another infant, even younger, was soon to arrive;

the pet baby named Pooh belonged to a young French couple who lived near the reserve and wanted to hand him over when he was old enough, but when at last he arrived, Stella herself had gone, to England.

She had been at home for more than a year since leaving school and felt it was time to start earning her own living. She applied for, and got, a job at the Wild Animal Kingdom at Woburn Abbey, in Bedfordshire. Woburn was the property of the duke of Bedford and was the first Stately Home (as they are called in the British newspapers) to open its doors to the public. The money made at the gate is important to such landed aristocrats, but what is even more advantageous is that the estate in this manner qualifies for educational status and gets a lot of tax relief. His Grace came of a family well used to the conservation of wildlife. His grandfather had maintained the first herd of wild Mongolian horses to be seen in Europe and had also collected Père David's deer when that species was threatened in China. Both horse and deer are now extinct in the wild, so it was natural for the present duke to improve his estate's attraction by creating a private zoo, as well as the more usual methods among his peers of conducted tours, gardens, gift shops, and cream teas. For a time he even ran a nudist colony, but that no longer exists. If I seem to expatiate rather a lot on the glories of Woburn, it's because it is only twelve miles from where I live when I'm in England, and I know it well. During the Second World War, when people weren't careful about keeping the gates shut, some of the Woburn animals got out of the grounds and made themselves at home in the surrounding woods. They (or their descendants) are still at home there. Out for a walk I have often encountered barking deer and other such creatures quite foreign to the country. Their bones may someday sadly puzzle paleontologists.

Anyway, Stella went to Woburn Abbey and was very happy there, nursemaiding young giraffes, coaxing them into the feeding area and anxiously guarding the public from their kicks. She took care of zebras. She took care of young elephants. She enjoyed it all, until the time came when one of the elephants was sold to a zoo and had to be crated for the journey. It came as a shock to face the fact that these creatures were not, after all, her property. Zoos had always been anathema to Stella, and now she asked herself whatever was she doing in one instead of taking care of her own chimps in The Gambia. She handed in her notice and went home, enriched by what most people would consider a most unusual experience. But then, women with apes *are* experienced in an unusual way.

She found that the colony now numbered eight apes, including Pooh.

He had been raised as a human child and was not now taking kindly to his surroundings. Stella's mother, having relieved him of his pink ruffled panties and other appurtenances of civilization, was doing the best she could with him, but he was scared to death of the other chimps and was unhappy, almost as unhappy, we might suppose, as a genuinely human child would have been. Stella did her best to reinforce his ego, treating him with special affection and helping him sympathetically to face the fact that he was a chimpanzee. Young chimps sometimes show their distress by rocking back and forth, and whenever Pooh did this, Stella picked him up and cuddled him. She had to comfort him whenever the other chimps snubbed him and wouldn't let him share in their games. At last he began playing with the others and broke through the barriers on his own account.

Watching the chimps, Stella made notes on their behavior: how William was braver than the others, how he and Cheetah had struck up a special friendship, and other such details. For example, there was the incident of the snake. Most chimpanzees are afraid of snakes, but William was not. Once Stella and her little troop came upon a dozing python twelve feet long, and all the chimps but William shrieked and cringed from the fearsome sight; but William approached the python and threw a palm-leaf frond at it. To be sure, the python went on dozing; but it was very courageous of William, and Stella was proud of him.

One day she got a bad shock. Tina, eldest of the females, was showing signs of maturity in that her genital area swelled up seasonally and became pink, at which times she accepted the amatory advances of Cheetah. This did not, of course, shock Stella, who knew far too much about natural processes to find fault with them; it was other behavior on Tina's part that she did not understand. Tina hunted and caught a baby monkey, bit its head off, and proceeded, in spite of Stella's protests, to eat it. Even worse was that the other chimps gathered around her and begged for pieces of the meat. We now know, thanks to Jane Goodall, that they were doing what comes naturally, but Stella hadn't been warned and was horrified, as if she had detected some terrible perversity in her chimps. Until Jane Goodall made her observations, chimpanzees were believed to be strictly vegetarian. Stella went on being distressed about it until a friend called to her attention the new Goodall book, In the Shadow of Man, in which the author told how Tanzanian chimps hunted down small mammals and ate them just as Tina did. What a revelation, what a relief for Stella! She learned many other things, too, and finally, as I have said, wrote to Ms. Goodall, who replied with a warmly encouraging letter. After that Stella was even more assiduous in keeping notes, copies of which she sent to the Gombe Stream Reserve, and the correspondence flourished.

Inevitably the chimps grew older and harder to control. Tourists began to complain of their behavior in the reserve. Stella had tried to teach them not to meddle with visitors' pocketbooks or steal hats, and as long as she watched them, they didn't; but as soon as she took her eye off them, they picked pockets, grabbed purses and hats and scarves and pens, and were generally naughty. Most visitors were too intimidated to grab back. William and Cheetah, the biggest, were the worst. Then William, while tormenting an antelope, got hurt when the animal charged, and Stella had to admit that the little group's days at the reserve were numbered, though there was a slight reprieve when the tourist season ended. Edward Brewer, too, found that they were becoming too obstreperous. They were always devising new ways to get out of their enclosure, and when they did, they destroyed many of his other projects—antelope enclosures, tree nurseries, that sort of thing. He told Stella that the chimps would have to go, and she sadly agreed. But where else could they be put?

Her father had a suggestion. North of the border between The Gambia and Senegal, about 400 miles from Akibo, was a national park, Niokolo Kioba, which he had recently visited and found to be beautiful. Moreover, it possessed an eminence called Mount Asserik, where wild chimps were rumored to live, though during his short stay he had not seen any signs of them. Did Stella think it was a possibility for their troublesome animals? Stella did. She thought hard about whether it was at all possible to transplant her colony there. Never for a moment did she assume that she could just take them there, and set them free, and go away and leave them. Chimpanzees are too human for such treatment; it would be tantamount to killing them outright. But if she went along to soften the blow and did things little by little, teaching them how to cope with nature in the wild, how to find food for themselves, how at last to do without her or any human being, if all these things could be accomplished—why, yes, Niokolo Kioba was a possibility. Of course, one would have to get the permission of the French authorities.

Her father's position made it easier than it would have been for a total stranger to get that permission, and A. R. Dupay, director of national parks in Senegal, readily granted it on certain conditions. Ms. Brewer would have to stay at a guard camp during the night because Mount Asserik was cut off at certain periods by rain and the rest of the park was not always safe. (Later this condition seems to have been softened, when Stella had proved that she could take care of herself.) We are not told if the Brewer parents objected to the plan. Probably they didn't. No doubt they figured that a girl who had spent a whole year unmolested in England could be trusted to look after herself in a wild animal park, and as things turned out, they were right.

Stella naturally wrote about these plans to Jane Goodall, who approved to such an extent that she offered a subsidy. She explained that Stella must estimate what the project would cost because, of course, it *would* cost; she had to think of supplies, camp gear, even a car if the arrangement was to be permanent.

First, however, Stella had to explore the possibilities in general. With a companion from her father's office, but no chimpanzees as yet, she made her first trip to the park. It was the rainy season, and, true enough, the roads to Mount Asserik were impassable; but she did look over the available park terrain and spotted several sources of food in the vegetation similar to those of Abuko. Encouraged, she went back to The Gambia to prepare for the move. Eddie Brewer agreed with her that they had better do it in sections. Cheetah, Tina, and Albert, the older chimps, would be the first to go because they were already accustomed, to some degree, to life in the wild—and were also the greatest nuisances. Cheetah was about eight years old; the others were seven, just about mature by chimp standards. Though William was only six, Stella revised her earlier ideas and decided he was big enough to accompany the others, mainly because he would miss Cheetah so badly if he had to stay at home. So William went, too.

Brewer was able to supply a Land-Rover for the trip. Stella, John Casey, one of Brewer's assistants, and an African driver were the humans. After a struggle they managed to coax the four chimpanzees into a traveling crate and set off. Keeping the chimps reassured and contented was very difficult, if not impossible. Tina's behavior was especially distressing to softhearted Stella, for she repeatedly took hold of the girl's hand through the bars and moved it toward the cage door lock in an urgent hint. The car was on the road about twenty-four hours because many stops were necessary. The next evening they arrived at the bank of the Niokolo River near the park boundary and made camp. In the morning they opened the crate and let the animals out.

The chimpanzees were ecstatic at being freed and rushed about hugging each other before they started to look around. But now what to do with them? Stella was too experienced to expect that they would immediately adapt themselves to freedom. They behaved much as she had thought they would, helplessly looking to her for guidance and food. William especially was adrift, but the other three also milled about for quite some time before Tina struck out along the riverbank, and the whole party, humans as well as chimps, followed after. Most of the time Tina seemed to have ideas about what to do. It was she who took the first drink from the river, and the others followed suit, all but William, who whimpered and would not drink until Stella filled a water bottle and gave it to him. Then he drank deeply. Tina also sampled one or two fruits, but even she was timorous and had to be taken care of. When night came, and wild chimps would have woven nests safely high in trees or bushes, all of Stella's band turned to her and slept as close as they could get to her makeshift camp. But Tina did at least know how to make nests, and after a night or two Cheetah and Albert imitated her and made their own. William, however, would not. He clung to Stella, and after several days she decided that it had been a mistake to bring him, that he was still too young. When John Casey took the car back to Abuko, William went with him. Stella, however, stayed at the camp near the river, the three bigger chimps still stubbornly sleeping close to her lodgings.

Standing by her agreement with M. Dupay, during the day she explored the park in the company of a park ranger or some other employee. She says she was never afraid, but at night she must sometimes have been lonely. Nights in West Africa are long. One usually goes to bed at sunset and gets up at sunrise. But nothing lasts forever, and soon the chimps were showing a gratifying tendency toward independence. Save for the occasional treat when Stella gave them a mango or something like that, they were living off the land. They had become a compact group, a sort of ape commune, and this, as Stella knew, is the natural way for chimpanzees to behave.

Day after day she wandered with her escort about the park, following her chimps. For resting during the heat of the day the chimps now constructed rather hit-or-miss nests; but at night, she was glad to see, they really took pains to build their beds, and they were now putting them properly high in the trees. Usually all the nests are in one tree. Stella was familiar with the various noises they made and knew in a general way what they meant. Now she noted a new sound, a throaty grunt something like their food noises but with certain differences. It was uttered when the chimps were settling down for the night, and she fancied it stood for a sort of roll call or good night signal. One hot morning she heard a warning hoot and looked out to see Cheetah standing at the edge of his nest like a sentinel, hair erect, staring at some movement at the edge of the clearing. The movement increased until an antelope stepped out from the vegetation. Tina and Albert now stood with Cheetah, and together they watched the antelope pick its way daintily across the clearing into the woods on the other side. The chimps' hair was in place again, and they relaxed. It was an encouraging sign, Stella thought, that the apes were alert to the presence of other wild animals. It was the same when the authorities set free in their neighborhood, without warning, eight unwanted lion cubs from a safari camp in France. (Lions are always in plentiful supply.) Tina, her hair on end, stalked toward the kittens until they ran away. But when a genuinely wild lion, a big one, came running by in pursuit of a bushbuck, the chimps stayed prudently hidden in the underbrush.

One gets the impression that there were few, if any, dull moments there on the riverbank. Tina caught a young vervet monkey. Albert carried off its head and ate it. Albert had begun to challenge Cheetah's exclusive rights to Tina when she went into her pink phase. Sometimes both males mated her, one right after the other, and though it had not yet come to downright confrontation, Cheetah was clearly not pleased. Ultimately it came to a fight, in which Cheetah routed Albert and Albert ran to Stella for comfort and safety. Then Tina's pink swelling subsided, and the chimps, all three, were friends again.

Five weeks after she had brought the party to Senegal Stella felt obliged to return to Abuko because her father was going to England and wanted to leave her in charge of the reserve. She was naturally reluctant to desert the chimps, but friends at the park promised to keep her informed of any important developments. Soon they reported that all four chimps had gone away together, nobody knew where. Later Cheetah came back alone to the camp and stayed there four days. The park people then transported him to a place near Mount Asserik and left him. There was no more news.

In her father's absence Stella might have found her responsibility overpowering, but soon she had help when a friend from Woburn days, Nigel Orbell, who had taken care of the bigger elephants, came to visit. The tourist season was approaching, and thanks to the departure of the bigger chimps, all kinds of preparations could safely be made in the reserve, such things as a new pool for the antelope, a blind for photographers, a bridge. And there was the usual collection of animal orphans. Nigel had to learn how to manage chimps, especially the rambunctious William, who in the absence of contemporaries was feeling his oats. During a run of bad luck they lost two of the remaining chimps, Ann and Flint, through pneumonia. Now there were three—William, Happy, and Pooh. Watching them, Stella worried. When the time came to take them in their turn to Senegal, she reflected, they would not be as well prepared as Tina, Cheetah, and Albert had been because they had always been surrounded by helpful humans, whereas the older animals had each had their own experience of living in the wild. Stella knew she would have to spend a lot more than five weeks this time, in a camp of her own, well supplied, and with a car of some sort. It would be expensive.

"Somehow," she wrote, "I had to earn some money."

Jane Goodall invited her to come to visit Gombe as soon as she was free to do so. She said that she had sent some of Stella's reports to her publisher, Collins, in London and suggested that Stella follow it up and get a contract for a book about her experiences, which meant advance money. Stella happily accepted the invitation, and as soon as her father returned, she went to Tanzania, where she stayed for two and a half months visiting Jane, Jane's husband, the photographer Hugo van Lawick, and the chimps of the Gombe Stream Reserve. She learned that from her venture she should have a stout car like a Land-Rover, and for the rest Hugo helped her to make a list of indispensables.

Still at the van Lawick camp, she had news from Niokolo. After a year's absence Cheetah had turned up yet again at the old camp, suffering badly from diarrhea. The people there fed him and gave him medicine, and he got better. On the fifth day he set out once more on his presumably lonely wanderings. But was he really alone? If not, where were his companions? Stella found it amazing as well as touching that he had actually found his way back, looking for help from his human friends. How had he managed to survive? She simply had to get back to Senegal, she decided.

From Tanzania, then, she went directly to London to collect money for her plans. Hugo joined her and helped. He knew the ropes. She soon had a promise from Collins for enough money to live on for a year, and Hugo added to her prospects by proposing to make a film of the project. Stella bought a secondhand Land-Rover and felt that at last she was really on her way. The mammal curator of the Regent's Park Zoo, Dr. Michael Brambell, proposed that she take two of the zoo's young chimps and include them in her rehabilitation project. The little apes, Cameron and Yula, had been born in the zoo and rejected by their mothers. They had never been out of their cages, and neither of them could be integrated into the zoo's chimpanzee group, so they were no good to the zoo. Stella wasn't sure the project would work, but she was willing to try. Dr. Brambell decided to wait and see what luck she had with her program. They would keep in touch.

These things were hardly settled when Edward Brewer telephoned from Abuko to tell his daughter that Happy was very ill and going blind, so Stella took the first available plane back to The Gambia, to find that the little chimp was indeed in a bad way and probably completely blind. A blood sample was sent off to England, but before the Brewers got the report, Happy was dead. The sample proved that he had been suffering from diabetes. So now there were only two chimps left, Pooh and William.

It was nearly 1974. Stella was resolved to go back to Niokolo Koba alone at first and try to find Tina, Cheetah, and Albert as well as establish her camp. Then she would return to fetch the younger chimps. With the family gardener this time, she did not go straight to the old camp but settled for a while on another part of the riverbank. The people at a nearby village had reported seeing chimps, and she wanted to investigate. They might not be her chimps, but she had to find out. Looking for them was tiring and frustrating. The humans found some deserted nests, quite fresh, and that was that. All the other nests they examined were old ones. At last they retreated to Niokolo Koba, and for the next two months looked thoroughly through the adjoining area, through the hills and valleys all over Mount Asserik. Once she saw a wild chimp, but it decamped as soon as it saw her. Having selected a good place for her permanent camp, she returned to Abuko just in time to get William out of her father's hair.

This time, to help with the two animals, Nigel went back with her to Senegal. After various false starts, which included meeting a troop of wild chimps that had no use for her, Stella learned that Tina had been seen at a place about fifty miles away, evidently on her own and following a group of baboons. They hurried to find her, and did so. After all this time of separation she was in splendid shape, and William seemed to recognize her instantly. At any rate he threw something at her. Then—for Tina was in her pink stage—they mated. Now what were they to do about her? Obviously she would be better off with her old friends than alone in the wild, but would she accompany them on her own? They borrowed a traveling cage, carried her back to camp, let her out, and waited. Tina stayed. For several days Tina, Pooh, and William moved about together, feeding and playing as they had done of old. One of the good things about having recovered Tina was that she could again serve as a model for the younger chimps, making her nest every night high in a tree. How to get the others to follow her example? Stella tried to be patient. Really there was nothing else to do. Tina did wander off soon after they arrived at Mount Asserik, but only for a few hours. After that she seemed quite contented to stay.

Nigel had to return to Abuko, but Stella engaged a young African to take his place. She and her new assistant built a platform in a tree near her camp, a nesting platform, as she called it. She reasoned that if the chimps would only build nests on this platform, they would be halfway to building them directly in trees. And though for some time to come they persisted in sleeping close to Stella, little by little things worked out as she had hoped. She now felt confident enough to write to Dr. Brambell, saying that her project was far enough along to include his little city-bred chimps.

Every day brought something new. Sometimes one of her group would get lost and scare her badly. Now and then a group of wild chimps drifted near the camp, ate from a fruit tree until the food was gone, and moved on. Once Stella, out with Pooh, came unexpectedly on three wild elephants. Both primates were awed into silence, and as soon as they could, they crept away without alerting the animals.

The odd little family had been living outside Niokolo for six months in their permanent camp. Slowly the chimps were growing more independent. It was October when Stella went into town to pick up her mail and there found a telegram from an unexpected source—Italy. A girl named Raphaella Savinelli wrote that she owned a chimp named Bobo, which at three years of age had become too hard to handle. Could she bring him to live in Stella's colony? Yes, said Stella, of course, she could. There was also an envelope from Dr. Brambell containing the earnest for an airplane ticket to London, where she was to pick up Cameron and Yula.

Raphaella, who arrived with Bobo in due course, was a competent, self-assured girl. She was one of those people we don't often encounter, who are good with apes as if by instinct. The first test for Bobo came when they drove back and presented him to the other chimps, always a rather ticklish procedure. Fortunately for the little animal, which seemed to have acquired his owner's sangfroid, he stood the test very well. He was too small to offer a challenge to William, and he showed no fear of the others, who accepted him very quickly, but the same cannot be said about Raphaella. William resented her immediately and tested her as soon as he could. When he saw her carrying Bobo, he bit her on the arm. It is true that chimp bites are not often as painful as, say, dog bites—their teeth are much like ours, and only the canines are pointed—but Raphaella gave back as good as she got, throwing things at him and running furiously toward him until he fled, screaming. Then Raphaella made peace with him, and for some time there was no more trouble on that score.

Nigel brought supplies and a telegram recalling Raff to Italy, which was a blow to Stella. Fortunately for her state of mind, Yula and Cameron now had to be fetched from London. When they got back, the fascination of watching a chimp deal with its native earth when it had never before been in contact with soil, let alone with free-roaming brothers, was absorbing. And the camp staff got another addition at the same time, an American girl, Charlene, who got in touch through Hugo van Lawick and who wanted to live there with Stella and observe the chimps. She was cheerful and strong, and Stella welcomed her help. It was Charlene who spent most of the time with Yula and Cameron during their early days in the camp. Unfortunately one day Cameron defied William, and William in a rage turned around and attacked Charlene. In what psychologists call displacement activity he bit off a piece of her ear, knocked her down, and nearly brained her with a bamboo pole. Charlene, badly shaken, went back to America, and again Stella was left without an assistant. As if this were not enough calamity, when at last Yula and Cameron were released, there was a donnybrook among all the chimps with Tina fighting on William's side. That morning Stella had taken the precaution of putting a few drops of Sernylan in William's drinking water, but at first it had no effect. At last the dope took over, and William fell asleep. Oddly enough, after he woke up, he was better-tempered altogether, and soon he and Yula were playing happily together. It wasn't a complete reversal of behavior. At least once more he lost his temper again and beat her up, but in general they were on friendly terms.

Raphaella returned. "Be careful of William" were Stella's first words at sight of her friend, and Raff, wise in the ways of chimps, handled the difficult animal perfectly. She calmed him down even before turning to the happy, welcoming arms of Bobo. But of course, this peace could not last. A few days later William attacked Raphaella, pushing her down so that she gashed her knee. She was angry enough to shoot at him with the camp's starter pistol; then she chased him and threw a big rock at him before taking the time to look at her wound. Her leg was bleeding, and William stayed prudently out of the way while she grimly sutured it without anesthetic. Later they made it up, and for a time the chimp behaved himself; but the peace was not permanent. One *would* like to hear more of Raphaella. After she had gone again, Cameron disappeared. They never found him, though Stella thought she heard his voice once.

Tina came and went as she liked. Once, however, she stayed away a long time, and her absence was explained when the African assistant rushed in with the startling news that Tina had come back, and with a baby!

Tina's baby was conclusive proof that Stella's rehabilitation project was, or could be, successful. As she finished writing her book, she could report that Tina had turned out to be a good mother and that William, presumably the baby's father, spent most of his time now with his mate and the little male, whose name was Tilly. All the others were there and accounted for, even Raphaella, who was now a full partner in the enterprise.

Stella's book first came out in 1977. A few years later two of its readers, Dr. and Mrs. Maurice Temerlin of Norman, Oklahoma, had reason to read it again with rapt attention, for Stella's commune gave them hope for their own lives, which were not ordinary. The Temerlins, as it happened, were the foster parents of a genius among chimpanzees, a female named Lucy. She is not the Lucy of anthropological fame in East Africa: that was a long time ago, in another country, and the wench is very dead. No, this Lucy was the intellectual heir to the first American Sign Language-talking chimpanzee from Nevada—Washoe. When Washoe left Reno and came to the University of Oklahoma and the special care of the psychology department headed by Dr. William Lemmon (under the extra-special care of Roger Fouts, who came from Nevada with her), a number of other chimps were put into training. Of these the star was Lucy, a baby chimp from the breeding farm run by Mae Noell in Florida. Lucy very early went to live with the Temerlins, both of whom were connected with Lemmon and the psychology department.

During the experiment other people adopted chimpanzees, but of these none held on to their foundlings as faithfully and stubbornly as the Temerlins. I happened to be in Norman at that time and visited several homes in which the infants were installed. It must have been rather like being a visiting inspector from some orphanage or adoption agency. I remember one baby that slept in a bassinet all filled with blue ruffles as well as the chimp, and another little ape sitting on a playroom floor, surrounded by toys and concentrating on a doll-size telephone. These little creatures, however, all came back in time to Lemmon's house and his many chimpanzees—all of them but Lucy, who grew and flourished at the Temerlins' and never saw another chimp. She was the pride of the teachers who made the rounds daily to tutor their pupils because she learned so fast and well. Soon she was being reported in the press for having mastered so many signs and for making up new ones, such as "candy drink" or "drink fruit" for "watermelon." But Lucy was also a problem, more so than even the chimps of Mount Asserik. She wasn't bad-tempered like Washoe (once described to me by Dr. Lemmon as a nasty little animal). No, it was just because she was so very, very clever. She would steal the keys of her room, hide them in her mouth, and let herself out when she saw her chance, to go rocketing around Oklahoma. She managed to start the engine in the family car and nearly drove off with it. She was more clever than most of the humans she had to deal with. Finally she had to be kept in a house of her own especially built over the Temerlin dwelling, designed expressly to hold this accomplished jailbreaker.

The Temerlins at the beginning had accepted the fact that their lives were of necessity centered on their foster daughter, but now they began to wonder where it was to end. Lucy matured, and Mrs. Temerlin telephoned me in New York to tell me that she had begun her period. "It gave me a strange feeling," she confessed. "I kept thinking, 'My little girl . . .'"

The Temerlins' natural son had grown up, taken his university degree, and moved on to a medical career. No problem there, but what of Lucy? Brilliant as she was, she couldn't go to the university. She couldn't even get married. The Temerlins tried introducing her to an eligible young male chimp from the Lemmon colony, but she was terrified of him and ran away as far as she could. (I saw a very telling picture of this: Lucy was sitting on a crossbeam up near the roof of her house, and Dr. Temerlin, who had gone up to soothe her, sat next to her, feet dangling, face perplexed.) Clearly Lucy did not feel akin to other chimps. To show her that they were not, after all, quite as bad as she thought, the Temerlins acquired a young female ape, Marianne, and introduced her into Lucy's house. This was successful; Lucy didn't mind such a small specimen of ape. But there the Temerlins were, with not only one chimp but another to worry about.

Now a strapping young female of eighteen, Lucy seemed, every time her surrogate parents looked at her, to reproach them. What would happen to her if anything should happen to them? Chimps can live a long time; Dr. Yerkes's Wendy did not die until she was forty-eight. The Temerlins would rather have liked to go to California and settle there, but with Lucy . . . Besides, they loved her. She was their daughter. Like Stella Brewer, like Joy Adamson with her Elsa, like others who have taken control of creatures from the wild, they could not contemplate life for their beloved animal behind zoo bars. Then they read Stella's book.

Clearly, it seemed to them, her colony was the answer. There Lucy and Marianne could be free—as free, that is, as any wild animal can be today in Africa. Like Stella, however, the Temerlins could not contemplate simply dumping their chimps in the wild and leaving them to cope, even in the company of Stella's chimps; Ms. Brewer herself would probably have to agree, and she might not like that idea. In fact, she held out for some sort of chaperon to begin with. A good deal of correspondence went on until, at last, Lucy and Marianne, accompanied by an observer supplied by the Temerlins, went to Senegal. The observer was to come and go as necessity demanded, but the chimps, of course, stayed. (And the Temerlins planned to move to California.)

Two years earlier, in 1977, Stella Brewer had married David Marsden, a forester working in Southeast Asia. She wrote to me about this in August 1984, explaining:

> Since I married I have spent an increasing amount of time out of the field and in the home, which is literally anywhere in S.E. Asia. . . . I go back to Gambia once a year to supervise running of the project. I take care of all administration and fund-raising now, which is a full-time job. Most of our funds these days are raised through an adoption scheme. People adopt one of 6 chimps and for an annual subscription they receive a biannual newsletter with color photographs of their particular chimp.

Madame Chiang Kai-shek, I recall, initiated a similar scheme for Chinese orphan adoptions during the Second World War.

Stella included in her letter a background article, or sketch, of the present commune or, rather, the commune as it was in 1982, when the piece was written. In 1982 she and her associates, as well as their charges, found it necessary to leave Mount Asserik. The chimps' numbers were increasing through births and additions; but the wild chimpanzees of the neighborhood had never become friendly, and now their attitude was becoming dangerously ferocious. Stella wrote:

> During the previous five years the camp chimps had struggled with the lessons of coping alone with daily life at Asserik and had succeeded. They knew their valley intimately, where to find the fruits in season, which areas to avoid for safety's sake, who were their neighbors and who would be enemies, where to find water and shade when the African sun was at its fiercest. The time was fast approaching when they would no longer need the security and guidance of camp and camp personnel, but at the eleventh hour, with the goal of true freedom and independence in sight, their very lives were threatened by the progressively more frequent and serious attacks from wild chimpanzees. I knew that even whilst we were there we could not protect them efficiently, and as soon as we left they would almost certainly be killed. Two months later, in two Land Rovers and trailer, a small caravan of four humans and eight chimps wound their way down the dirt road away from Mt. Asserik.
>
> Our new home is part of the River Gambia National Park in The Gambia. We are constantly being asked to take on more chimpanzees. Between us all, Janis, myself and my father, we are now responsible for the lives of thirty-six orphaned chimpanzees.

This, remember, was in 1982; no doubt there are more by this time. And there is one more Marsden as well; Stella has had a baby.

Janis? That is Janis Carter, the observer first supplied by the Temerlins who came out to Africa with Lucy and Marianne. The names of all three of them are included in Stella's list of the chimps on what is now called Island II. Janis's story is told in Eugene Linden's book *Silent Partners*. He describes her as a plump, independent and legendarily stubborn woman who in the mid-1970s went to the University of Oklahoma from the University of Tennessee especially to work with the signing chimpanzees of Lemmon's colony. The connection did not last long. Though she enjoyed working with chimps, she took exception to Dr. Lemmon's methods of maintaining dominance over them and soon stopped the language lessons in favor of helping the Temerlins with Lucy and Marianne. From the first Janis and Lucy got along well, and when at last the Temerlins made up their minds to repatriate the chimps, if that is the word, they consulted Janis as a matter of course. She was more than willing to go along and help the chimps in Africa, and her inclusion in the scheme tipped the scales with Stella, who had been unwilling at first to accept a fully grown tenderfoot like Lucy. Janis, the Temerlins, and the two apes left Oklahoma for The Gambia in September 1977. We cannot be surprised to hear that Janis was very doubtful at the beginning. As she told Linden, she wasn't used to the sort of life she would have to live in Africa; she didn't even care for camping. As for her charges—"The idea that Lucy, the fastidious, toilet-trained chimpanzee princess, could make it in the wild stretched the odds to the limit," wrote Linden in *Silent Partners*.

Certainly the beginning of the venture was daunting, and the Temerlins, who stayed only a week, must have felt in an unhappy state of mind, tormented by guilt and worry. Janis did not fare much better than the chimps at first, and they suffered sadly, confined to cages as they were. The Brewer reserve at Abuko was supposed to be a temporary stop for them before being moved to the Baboon Islands, and because if Lucy were permitted to run free, she would never allow herself to be caught again, she stayed cooped up, moping and getting thinner and thinner. Janis, herself not much more comfortable—she was living in a tree house nearby—tried in vain to persuade the older chimp to eat native fruits and berries; Marianne, being younger, was more adaptable than Lucy. At last Lucy did start to eat something, egged on by the sight of the young Dash, a male chimpanzee who like her was awaiting deportation, swallowing the food.

It was a long time before Edward Brewer could be persuaded to move Janis Carter with her two animals, along with Dash and six others, from Abuko to one of the Baboon Islands. It was May 1979 when they did so. Stella has described this new venue in her book. It is not at all like Mount Asserik, being flat, green, and densely vegetated along the periphery. Two miles upstream from her island is Janis Carter's Island II with her group of nine chimps, "new to the world of foraging and freedom," as Janis describes them. Downstream 180 miles were waiting, at the time of publication, eight more chimps, all but two of them infants. There were too many chimps looking for too few homes.

"We have run out of space in The Gambia and have to establish a new center somewhere before we can accept any more chimps," wrote Stella.

Eugene Linden went to see how Janis and Lucy were getting on and found Lucy looking tall, thin, and distinguished-looking. The bond between the woman and the chimp was obviously very strong, even though, as she reported, Lucy could probably manage without her, as far as her life in the jungle was concerned. Janis (who lives in a cage while the chimps roam free) reported that the great breakthrough took place when Lucy first opened a baobab fruit herself. Until then, though she was capable of doing so, she had always handed it to Janis to open.

Janis lives in that cage and no longer contemplates leaving Island II permanently, though she has made short visits home from time to time. Which primate is the prisoner? All we know so far is that Lucy is dominant in her group, but then, she always was.

8. *B*ARBARA HARRISSON: MUSEUM DIRECTOR

"I STARTED WITH THE ORANGUTANS when I was about thirty-three and went on for about fifteen years with that work," said Dr. Harrisson. We were in her office in Leeuwarden, the Netherlands, where she holds the position of director of the famous Princessehof Ceramics Museum, a surprising change of direction for one who used to be described, feelingly, as the scourge of all zoos that exhibit orangutans. Our acquaintanceship was short if one judges such things by personal contact, but for some months we had been corresponding. I had first heard of Barbara Harrisson something like twenty-five years earlier. Visiting Sarawak in Borneo, I encountered her husband, Tom Harrisson, who was running the Sarawak Museum in Kuching, Sarawak's capital city. She was not in Kuching at the time, but off somewhere in the forest, observing the apes. In 1963 she published a book entitled *Orang-utan* in which she told a good deal about the subject, the nature of the apes, and how they fared or did not fare in zoological gardens; she was bitingly critical of these institutions, I recalled, and the book showed her as an unusually intelligent observer with a detachment rare in such commentators. Not that emotional involvement was lacking; far from it.

When I heard of her presence at the Ceramics Museum, I sat up. None of the other women in this study have given up their fascination with subhuman primates or at least have journeyed so far from the animals. How had it happened? How and why had she broken away? Therefore as soon as I heard where she was and what doing—and it was a purely accidental discovery; my husband, a ceramics buff, met her when he visited the museum and mentioned her to me—I sat down and wrote her a letter asking these questions as well as her theories as to why some women should be so successfully involved with apes. She replied immediately, though as she explained, it was a very busy time for her: She was moving house. She said my questions were interesting, so much so that they could not be answered by letter, and if I could see my way clear to coming over to Leeuwarden once she had moved. . . ? Yes, of course, I could. So, on a day that was snowy and bitter cold, I landed at Schiphol Airport near Amsterdam and took the train to Leeuwarden. It turned out to be quite a major project, going from Amsterdam to this town in northeastern Friesland—two hours at least. But plenty of people make the trip, if only to see Barbara's museum.

"I knew nothing about apes and monkeys before I came to Sarawak," she began. "Working with these animals began because I had to look after some of the young ones at home."

She had to learn the hard way. By the time she got to Borneo the orang was already classed as an endangered species in its native haunts, Borneo and Indonesia. In Borneo, especially in British-ruled Sarawak, the authorities had been prevailed upon to pass certain conservatory laws. To get possession of a baby orangutan, the age preferred by collectors because large orangs are so strong, hunters had to kill the mother. Now it was illegal to buy any of the little creatures or to keep one as a pet, a rule hard to enforce for several reasons. People who had lived in those parts all their lives—natives such as the Dayaks, Malays, and Chinese—probably did not see why they couldn't do as they liked with their own animals. And dealers paid well, according to native standards, to get hold of the creatures. Besides, the orangutan in its infancy is very appealing, humanlike with its red hair. Mature animals are difficult to control because they are immensely strong and very clever at escaping, but the babies are different. People of all races liked them. So the law was disregarded by many residents, and it became necessary to confiscate any little orang that officials happened to find in private possession. Only a few people were officially permitted to keep orangutans for the purposes of conservation. And one of these privileged persons was Tom Harrisson, who by marrying Barbara made her a sort of nanny for all the infants collected and brought in by the forest guards. As curator of the museum Harrisson had official standing and served (willingly) as *the* expert on apes they found and turned over to him, rather than the governnor of Sarawak. In any case, one could not imagine an orangutan in Government House.

Barbara is a tall, handsome woman who speaks English from which all trace of her German origin, except for watchful enunciation, has been eliminated. (After all, she has to speak Dutch all day.) We sat in her new house in a village just outside town. I asked about her training in primatology.

"I have no degrees in that subject," she said. "The degrees I hold were earned later, after Borneo—a Master of Arts and a Doctor of Philosophy in art history—and I got those degrees at Cornell. I do have an honorary degree in science from Tulane University for my work with orangutans, but this was only after I had finished with all that. I don't remember how many animals I looked after in Sarawak, but you can find all this in my book."

She was right. I looked it up in the index. Fifty orangutans are mentioned in the text, each by its own name.

"I found it rather an interesting thing to do," said Barbara, "I guess partly because I had no children, and these orangutan babies are nice if you have them under a year old, because they very much resemble human babies of the same age. The trouble starts when they get older. You have to go on treating them like babies, but at the same time they are very destructive and can get very strong. You have to cage them, and then you are in the business of running a little minizoo, next to your home. I do not like that because I like to see animals in nature. Essentially I don't like caged animals."

She spoke with firmness and as if she had often had to explain her feelings in the past. "So then I got the idea of trying to put them back," she continued, "where I thought they really belonged, and this was the start of this experimentation."

It started on Christmas Day in 1956, when Tom brought Barbara, who was lingering in her bedroom with a cold, a baby orangutan like "a ball of chestnut red fur." She described it on page 2: "It was a beautiful baby with wide dark-brown eyes in light-colored sockets, a small nose like a triangle with a velvet skin and a broad mouth with soft sparse hair like a young boy's. His body and long arms and fingers, his short legs and long toes were covered with hair, hard and shaggy. On top of his head this hair was sparse and upright, framing the face and small, human ears with a halo of sweetness."

Tom told her that the little male's mother had been shot, that he had been found by a forest guard in a Dayak longhouse and been confiscated according to the law.

"We shall have to look after him for a while," he said.

A friend named Bob was coming to dinner, so they named the little ape Bob.

As is the nature of orphaned infant orangutans, Bob immediately adopted Barbara as his mother and clung to her with hands and feet, screaming if she pulled him off. Within a few days it was obvious that some new arrangement must be made for the little newcomer, someplace for it to stay when it wasn't hanging on to its mistress's clothing. Fortunately the Harrisson house was adaptable. Kuching was then a town of forty thousand—Chinese, Malays, Land and Sea Dayaks, Indians, British, and anybody else who happened to have taken a fancy to the place and settled in. It was not difficult to take a fancy to Kuching. "A dream-town of the old eastern style, with a Chinese core," as Barbara put it. The Harrisson house was in Pig Lane. She described it:

> A narrow passage is lined with Dayak swords and hats, crammed over black, white and yellow monsters painted on the walls by up-river visitors. Coming out into the main room—a vast open veranda looking over the garden with low rails allround—lady visitors of orthodox tastes often exclaim, "How can you *live* here?" Others, at best, remain silent, scanning the room with awe. For this is the housewife's bad dream, conglomeration of things Bornean, pinned and stuck on wall and ceiling, lying on tables and floor, everywhere and nowhere: war hats and feathers, shields and swords, old bead necklaces, shells and dragons, statues, Siamese buddhas, plates and wooden carvings, drums, ceremonial sticks, Ming jars, masks, baskets, mats, *tapa*-bark boats, fine sarongs, Rhino horn, hornbill ivory, Birds of Paradise, and on every free inch of table piles of books and papers, neatly arranged by their master in sections of planned activity and interest where no one may touch them, even to dust.

The bedroom was equally crowded; so were the study and everywhere else that had been added to the dwelling. In other words, a curator's house—a curator with interests in archaeology, anthropology, ethnology, and sociology. Somehow Bob had to be fitted in. The Harrissons simply added a cage in a spot not too sunny and not too wet. When Barbara took Bob to the local veterinarian for a checkup, the verdict was a healthy infant on the whole, though he had a slight cold. He weighed fifteen pounds. His milk teeth were fully developed, so Tom decided he was about a year old.

"We must weigh and observe him regularly," he declared. ". . . I will give you some cards, and every day you observe something special you write it down and keep it on file. But you must *really do it*, don't just say 'yes.' "

She did really do it, as her later writing attests. She became a trained scientific observer with high standards. But as time went on, it became increasingly impossible to be Bob's mother, Tom's wife, and a museum worker all at the same time, so they hired Bidai, the fifteen-year-old son of a Land Dayak who wanted his son to learn city ways. Bidai took over the care of Bob, cleaning the cage, spoon-feeding him with milk, and playing on the garden lawn with him for hours. In those days it was an easy job, but soon there arrived another baby orang, a female they named Eve. Another cage was put up next to Bob's, and Bidai had a new charge. Eve was tiny and starved, weighing only seven pounds at first. The forest guard had found her in a Dayak longhouse. She had been kept on a chain by her first owner, a Chinese trader who had allegedly bought her from hunters, and she was in a poor way, her neck badly chafed by the chain. She would eat nothing but small bits of banana, refusing milk until Barbara shoved some into her throat with a pipette. Then she rapidly learned to drink it and soon gained weight. Bidai had orders to put fresh branches into the cages daily because orangs like to eat leaves and pull the twigs about.

Months passed, and the apes flourished, though orangutans are very slow growing. Eve became spoiled; Bob remained as he had always been, contented and good-tempered. Now he was three years old, and there could be no question of keeping an adult orangutan in a house, even the Harrisson house.

The conservator of forests, who was also the head game warden, admitted that the confiscated infants were becoming a real problem. There was no possibility of keeping Bob indefinitely in or near the house, even—or especially—the Harrisson house, with its many artifacts. For some time Barbara had been corresponding with a woman who was a keeper at the San Diego Zoo and whose special charge was a young female orangutan named Noelle. They had compared notes on the proper care and feeding of orangs, then in very short supply in the zoological garden world, and from what Barbara gathered, San Diego was an unusually good specimen of its kind. Was she actually thinking of putting Bob into a zoo? Yes, she was. It was a question of temperament, she told herself; Bob was not the stuff of which pioneers are made. In the end she wrote to San Diego offering him as a bridegroom for Noelle, who was about the same age. The authorities in

California were, of course, delighted to accept. But before Bob's dispatch could be considered, let alone arranged, Pig Lane became home for three new baby orphans, all males.

By this time there was an unwritten agreement between the forest department and the museum that confiscated orangutans were to be given to the Harrissons and they would bring them up. "No one else was interested," wrote Barbara, "so we had to start devising mass methods of raising them, now. Perhaps it was even possible to find new ways of keeping them, in a half-wild state, so that they might be put back in the jungle later? If this could be achieved at all, it would be a long-term project."

This was the beginning of a number of the attempts that we might with justice describe as worldwide, to restore wild animals to the wild. It had been attempted in a small way with lesser animals, especially birds, but nobody had yet entertained the idea on a large scale with such creatures as orangutans—not that there are many such creatures. From that time on, Barbara was even more scrupulously careful to keep account of the animals that passed through Pig Lane, as they arrived, lived and were taken care of. Tony, Frank, Bill, and many more—there they are on record with their weights, diseases, special characteristics, and fates. Barbara, true to her promise, kept faithful records of all of them. Thus we can learn that Bob was sent, eventually, to San Diego via Singapore. Wrote Barbara, "Tom could not bear to see him off," but he felt better when a letter came from Dr. George Pournelle, San Diego's curator of mammals, with news of the animal. Bob proved to be so good at escaping from cages that they considered renaming him Houdini. Finally, however, he settled down happily with Noelle.

One day in 1958 Bidai neglected to supply enough fresh branches to the caged orangs in Pig Lane, and Tom was furious with him.

"You must get more branches," he stormed, and when Bidai excused himself by saying that there weren't many left within his reach, Tom retorted that in that case he must climb higher. Bidai did, and Barbara noticed how wistfully the orangutans watched him going higher and higher. She mused, "Perhaps we were wrong and should give them more freedom. Perhaps they should learn to climb high and live in the trees. Was it possible, in this way, to bring them back to where they belonged, in a natural life, once they were strong enough to look after themselves? But when was this? Bob, at three (and fifteen pounds) had still been a playful baby, though strong. Was he, at that age, depending on a mother to teach and protect him in the jungle?"

There was no book to consult about the matter, no field report. The Harrissons would have to find out for themselves, and this meant fieldwork. But Tom couldn't drop his museum to go out chasing apes. It was up to Barbara.

The museum had in its employ a man named Gaun Anak Sureng, its star collector and taxidermist, a keen hunter who had done a lot of traveling in the jungle. Tom sent him out alone, at first, to find the best areas for observing orangutans in the wild. Then Barbara prepared to go out with Gaun on a second survey. She collected the necessary items for such a trip, which would be rugged—cot, mosquito net, food, changes of clothing—and the little party of four—Gaun, Bidai, a friend of Bidai's, and she—set out for the Sebuyau district, where, according to report, wild orangs had been seen. They traveled by boat for three days, and the next night after that they made camp in a longhouse. Walking through the secondary, cutover jungle as they had to do was tiresome, though the men cut away undergrowth and when necessary gave Barbara a hand across slippery log bridges. They saw no orangs that first morning.

A sure sign of the presence of the apes was orangutan nests, old or fresh; it was for these that Barbara kept looking. In an article she published in the *Sarawak Museum Journal*, she went into detail on the nests. In her introduction she wrote:

> All Pongidae [i.e., the great apes] build nests, but chimpanzees and gorillas give birth on the ground, while the record for the orang-utan remains uncertain. . . . The fact that Pongidae construct nests is assumed to have evolved in the arboreal proto-apes as a behavioural adaptation to gradually increasing body size and weight. (Reynolds, 1967). But fossil proto-apes have been found in an open plains situation and there is no evidence that proto-Gorilla and proto-Chimpanzee were not just as big as the recent form (Leakey, verb. comm.). Whatever the theory of evolution, it is exceedingly hard for arboreal creatures over a certain size to find a situation in a tree to accommodate it securely in sleep unless a nest or supporting platform of some kind is built. For large tree-sleepers nest-building becomes a daily requirement, the construction mainly functioning as *sleep-nest*. The endogenous stimulus for sleep-nest construction is somnolence, triggered in the nervous system and automatically taking effect in the presence of external stimuli—such as a suitable site location, nest-building materials and other environmental criteria.

Later in the article, which was also privately published as a brochure, ("The Nesting Behaviour of Semi-Wild Juvenile Orang-Utans"), she writes:

The animal usually starts the activity from a quadruped position. If nesting in an arboreal environment, he sharply bends or breaks leafy branches (sufficiently strong to support his weight) towards himself with his hands. He holds and presses them with their external ends underfoot, so that they come to rest between himself and that fork or tree limb he had selected as his nest base. After breaking and bending them into a rough platform, he presently sits on the tangle to continue incorporating smaller branches and twigs, broken or nipped off with fingers and (or) teeth. He lifts them slightly overhead, turning them "outside in," so that they come to lie in his crotch with the terminal leaves pointing towards himself. He compresses the material by giving it pats with the back of the cupped hands, as if "drumming." The patting may be followed by a stronger downward pressure of hands, supported by a forward bending of hands and shoulders, a simultaneous straightening of arms and slight lifting up of the posterior, so that the animal's full weight rests fractionally on both fists. Then he sits down, slightly turning round his axis in doing so, to proceed picking, now from a different angle with new branches in his view and reach. No weaving, plaiting or knotting of nest materials takes place. The animal may use the nest right as it is, after construction, or he may try it crouching, squatting, sitting, or lie in it briefly to work some more afterwards. He may get up and climb a distance to bring back further branches from elsewhere, or he may abandon the nest and start all over again, at a different spot.

On the second day the party found no nests and no animals. But on the third day they spied an old nest, and then, about midday, they actually saw a fresh one, twenty feet up in a tall durian tree. (The evil-smelling durian fruit is very popular with orangutans as well as with the natives of Southeast Asia. Raffles wrote that it smelled of drains. Yet if one can ignore the smell, it is a good, refreshing fruit, tasting something like sherbet.) Barbara was just taking a photograph of that nest, the first, when a big orangutan got out of it and departed in a temper; Gaun said that they had disturbed the animal's afternoon nap, so it might not be tactful to follow it just yet. However, they saw three more, young ones, the same day. Following them, camera at the ready, Barbara didn't watch where she was going and, to her embarrassment, stumbled and fell down. An orangutan peered at her prone form, dropped a twig on it, and moved off without haste. Later they found the first orang they had seen and watched it improve its nest, bringing in new branches, tucking them comfortably around itself, and finally settling down to sleep in spite of the rain that had begun to fall. Watching, Barbara thought guiltily of the orang babies at home in Pig Lane, protected and cuddled snugly into bedding in their cages. Could such cosseted infants possibly learn to cope with life in the jungle? She doubted it.

"The babies that had come to us were in their helpless stage," she wrote in *Orang-utan*, "when they still entirely depended on their mothers. Being adaptable and intelligent animals, they had accepted the human being as suitable substitute mother and started to learn and live in a *human* way. Was it humanly possible for humans to teach them the *ape* way—so that they might be free to return to tree and jungle life? . . . I could not help being doubtful."

She came to the conclusion—it was a restless, thoughtful night—that unless and until she learned to live like an ape, it would be impossible to teach an ape baby how to do likewise. At last she slept, but next morning, very early, she determined for a start to get into a tree herself, if only to photograph the young orangs. She found the tree where the three young-sters were nesting, and Barbara, with her camera, got up onto a branch while the young apes were still asleep. She was there, watching, when they woke up, found ripe durians to eat, had their breakfast, and fell asleep again. (Wild apes tend to be admirably relaxed.) She was still there when they awoke again at ten and began to get as busy as wild orangutans ever get. One was making a new nest, and Barbara was just wondering if she would ever be able to walk again on her dead legs when an ape wandered quite close to her, and Gaun on the ground fired his gun to warn it off. All three apes immediately disappeared very speedily, and she came down at last to Mother Earth.

The party got back to Kuching three weeks after starting out. Yes, reflected Barbara, it had been a good experience as these things go, but she was no wiser than she had been before as to the social life of orangutans. At least, however, she could now be a true ape mother. The children in Pig Lane must spend more time in trees, possibly even making nests. Which of them had the temperament to embrace this reform? Eve was still a prima donna, clinging to Bidai, reluctant to climb anywhere. Bill was more interested in eating than anything else and was getting rather fat. But Frank quite obviously loved to climb. He would go as high as he could, with Bill a slow second.

Because of Eve's hysterical attachment to Bidai, Tom decreed that the boy should go on leave to his native village. Eve did not seem to mind his disappearance, but she still wouldn't climb unless forced to do so, unlike Frank, who could only with difficulty be tempted to come out of his tree for dinner and at night. There it was, Barbara decided. Frank was older when he was orphaned and had already been coached, at least a little, in jungle life. The Harrissons heard that the Berlin Zoo had a new ape house. Eve was certainly better suited as a zoo animal than as a candidate for freedom, so Barbara wrote to Berlin offering her, and the zoo authorities were happy to give her a home. She went off to join a group of young orangs in Germany, but the Kuching apes soon became a threesome once more with the arrival of Nigel at the end of November 1958. Nigel was unusually large for an orphan, being about two years old, and as might be expected, he was far more independent than the younger infants. One day in early December, after spending the daylight hours in the museum trees, he wouldn't come down for his evening milk, but at six o'clock he began something not seen before in Pig Lane: he built a nest. This was very exciting to his guardians, as well as impressive, because it proved that he had not forgotten his natural ways even though he had been imprisoned a long time and at an early age. That night Nigel stayed in his tree until daybreak, and he was slow to come down for breakfast milk, though he was obviously hungry for it. He did descend at last, had breakfast, and went to his cage for a good long sleep. That evening he began to make another nest, and Frank, after watching for a while, tried to help him, bringing twigs and branches, which Nigel nonchalantly accepted. Later Nigel started on another nest higher up in the tree.

In the course of the next twenty-four hours he refused to come down at all for milk but ate the fruit he found in the tree. All this, of course, was tremendously interesting to Barbara. Nigel seemed quite self-sufficient when he had to be, she reflected. During the following fortnight he spent more time than not in the trees, though once he did deign to descend and sleep in the cage on a rainy night. Had he foreseen the rain? At any rate he was the answer to some of her questions. Through all this, Bill showed little or no interest in the wild, or half-wild, life, but Frank was more and more drawn to it. However, at last even Bill began to fool around with branches and, in a halfhearted way, make a nest, but his main interest remained food. In the end, however, Barbara had to admit that none of the orangs showed much interest in being wild as long as they were fed. Orangutans wander through the jungle chiefly to find the fruit and leaves on which they depend, and if these are supplied, why should they wander at all?

Little by little, as the fruit was eaten from the museum trees and disappeared, even Nigel stopped moving around very much. The Harrissons agreed that something new would have to be done if they didn't want to produce a whole generation of animals fitted for zoo life and nothing else. However, the orangs still swung and jumped when they felt like it—even Bill did that—and the trees, overworked as they were, looked pretty bad. Barbara relaxed the boundaries of the orang territory, and when they had ruined the rest of the museum trees, she got permission to take them over the road to another estate, where they set to work happily ruining the trees there as well.

Clearly the apes had learned something of life in the jungle, but they needed more rigorous training, for there would not always be a cage waiting for them to sleep in at the end of the day. What was needed now, Barbara felt, was less comfort, larger territory, and enough food around for them to find it for themselves. How to get these things? Put the orangutans into a sanctuary? Where? Anyway, it would cost a fortune even if there were such a place. Obviously it would be best to turn them loose in the wild, but there were difficulties in this plan. An orang that was used to people would be too friendly and simply ask to be killed or captured. Also, a liberated ape should, of course, come into contact with wild orangs, in order to adapt to their life, and in these days that isn't so easy, because wild orangutans are disappearing at a frightening rate. The third difficulty, Barbara explained, is that the orang learns its roaming propensities by the age of two, from its mother. The little animal gets its main education between two and four, and if that time is spent among humans, it becomes humanized and doesn't learn to cope with forest life. She wrote in the *Museum Journal*:

> The slower an animal develops the longer is the teaching it requires from its elders. It is quite wrong to assume that a young Orang reacts to a large extent "by instinct." In my view the contrary is the case: most things are learned through direct teaching, and many fundamental attitudes—climbing trees for instance—are quickly forgotten by babies if they grow up on the ground to the extent of being frightened of any real height. . . . The Orang's natural curiosity and keen sense to explore, his slow contemplating mind with a

capacity to remember events, makes [sic] him *a priori* highly successful in the jungle. But these very attributes also make him successful, especially as a juvenile, in *human* surroundings, as is the case with our children at Pig Lane. This shows the importance of education.

Apart from all these calculations, she added, there is emotion. How can anyone who has loved such animals bear to put them back behind bars? Yet it is the only alternative unless one is prepared to abandon everything else and build one's life, home, and economy around these animals as they slowly mature "to enormous strength." The dilemma is dreadful, wrote Barbara, and she compared it with that facing many Europeans she saw around Kuching, who had to send their children home to boarding school, sometimes not seeing them for years at a time. In spite of her feelings, however, the Harrissons felt that they had to give yet another little ape to a zoo; at the end of March 1969 preparations were completed to send Bill to the Antwerp Zoo. He was the one best suited for captivity. "He had been self-satisfied, of balanced temperament, from the very beginning," explained Barbara. "His response to half-wildness had been slow and rudimentary; he seemed to prefer an easy cage-life with plenty of food."

But, as always, someone came to take his place. This was Ossy, a very young ape that Barbara found herself unable to care for as he needed, because Tom was away and she was in charge of the museum. Ossy was farmed out to the veterinarian's wife and little girl, who adored giving him his bottle. Then came Janie, a female that weighed only five pounds and did not live long. When Barbara had more time, Ossy came home, and she was able to observe him closely, learning that he could make out newspaper photographs and the flowers and leaves that were printed on her dress. Frank and Nigel were still living their half-wild existence, but they had exhausted the possibilities of all available trees and were slowly settling in to a pensioner existence, taking the food handed out to them and sleeping all night in the cage. By February 1960 both of them were large and flourishing and an increasing problem. The Harrissons were going on leave in May, so dreams of planting orphans back in the wilderness had to be postponed. In the meantime, the Hamburg Zoo declared itself delighted to take the boys, and in the Edinburgh Zoo the director was so eager to obtain Ossy that he promised to build a new, special house complete with glass sun porch for the little creature.

It remained only to get the animals to their new homes, but this was not easy. Tom didn't have the time to go home to England by boat; he would fly. It was up to Barbara to chaperon the orangutans, but before she could even begin to arrange passage, another infant was brought in, a young male they named Derek, just the right age to be a companion for Ossy. Of course, he went along, too. In the middle of May Barbara, with four orangutans in two cages, set sail in the only ship that would take animals as cargo. It was, to be brief, a difficult and long trip; there were storms, the cages had to be moved around to avoid flooding, and it grew very cold.

Tom met them in London and flatly refused to consider Barbara's proposal that she herself take the two older animals to Hamburg. They were sent in somebody else's care, and the two infants were picked up by an emissary from Scotland. He and his wife had been separated a long time, declared Tom, and he wanted her back.

"I felt defeated," wrote Barbara sadly. "In spite of all efforts and understanding, I had not done better than to expose [the orangs] to an uncertain fate in a zoo."

The rest of the book is given over to her thoughts on zoos and her reports on some of them, which she later visited and studied. The Introduction was written by Tom and includes the following passage: ". . . though I love [my wife] as life itself, there have been times, in recent years, when I have wondered who was living with whom. She with me—or Derek and Ossy? I with her—or with Nigel at the shaving soap?"

Ultimately, it seemed to Barbara, it came to a question of choice: her husband or the orangutans? But the moment had not yet arrived when the Harrissons got back from leave, and orphans continued to come to the museum. In June 1962 it seemed as if she were able to satisfy all demands by putting the apes in Bako, the state's first national park, started by the British administration in 1957. Bako is twenty-three miles from Kuching, and its area is about ten square miles. Unfortunately it held no population of free-living orangutans, though there were plenty of monkeys; but at least Barbara could get there and back without losing much time, so at first Tom did not resent the apes' claim on his wife's attention. However, as Barbara tells in another article she published in the *Journal*, this area was not a good enough place, and soon she moved her three juvenile orphans farther away, to a locality in the park called Telok Asam. This was the main administrative land of the park, a beach camp in the west. Unfortunately it was easily accessible to tourists, who came up from the beach to look around and did not take kindly to being hassled by orangutans, however juvenile. They complained. After nearly a year of this Barbara in April 1963 transferred the orangs to Telok Limau in the north, where visitors were few. There the apes stayed for two years, when they were moved yet again in what was to be their final peregrination, to the east coast of Sabah, 800 miles away, into the Kabalit Sepilok Forest Reserve, where they were incorporated into a new rehabilitation project modeled on Bako and set up by the Sabah Forestry and Game Department the year before.

"The animals were looked after at Bako, and that was very well," Barbara told me, thinking back, "but there was no continuity in that they could make contact with a wild population and integrate there. . . . Because the animals had no fear, they came down from the trees and harassed the tourists, and that was why I had to relocate this project. Just before leaving Borneo, in 1967, I succeeded in transferring all these animals to Sabah. This is a project that the Sabah government continued, and it is still in existence. But something happened that put me off: The largest male I had, who was, I think, about eight years old at the time of capture and was very strong and rather aggressive, was shot by one of the gamekeepers there, a man who I think acted to spite me and also out of fear. And this I found very difficult. It happened while I was away, but it really brought home to me the dangers of doing such things and the negative aspect. Anyway, long before I transferred the animals, I had been in conflict partly with my other, private life. Every time I wanted to stay more than a week, say, to do something with the animals, for what Biruté Galdikas and the others are doing right now, I felt I had to go back in order to pick up my private life. This had probably to do with the fact that Tom was a very demanding person, though very amusing, very positive, and very interesting to live with."

The difficulties she had with her husband seem a pity. Barbara Harrisson was important in the history of wildlife conservation. One has only to read the *Journal* article to appreciate the value of her contributions to the cause of rehabilitation; it would not be too much to say that she set a pattern that has been followed by everybody who came into the field after her. She thought it out and planned it on the foundation of her unique knowledge of the subject, and it is admirably presented.

As she said in another connotation, "Notes. You keep notes all day and, if necessary, all night. Of course when the project is finished, you have far too much material, but you *must* go through all the notes again. That is the hardest part of the game."

"It was a stimulating life," she now continued, "but Tom was a terrible chauvinist. However, I was married to another man before and that had ended in divorce, and I was determined to make this work. Somehow I had to make a decision: What do I want? To break away and simply go into the forest and do this orang experimentation, or do I wish to continue this other existence and do less about orangutans? Well, I chose to do less about the orangs for two reasons: one I have already explained, and the other because life with Tom was stimulating and I thought it rather nice. Furthermore, a negative aspect of my life with orangs was that I did not have a degree to do it properly, and I would have found it difficult to get funding. They want you to have a master's at least.

"I spoke to Louis Leakey about this when Tom and I went to East Africa, to Kenya, on one of our leaves. We were invited to stay at Government House, and this was very interesting because Malcolm MacDonald was there as the last British governor; it was at his invitation that we were there. Because we had this relationship with him, Louis Leakey started to look after me while Tom was doing the archaeology bit with Mary Leakey and the others. Louis, of course, was doing the animal bit with me. (I was interested to read the other day that Mary has written a book about Louis. You might consult this as to his relations with women.) He talked about Jane Goodall. . . ."

Ms. Goodall's first contact with chimpanzees, as the world knows, was due to suggestions made by Leakey, and she continued for a long time to consult with him after her project was firmly settled in Tanzania. Louis Leakey, as she has written, helped her find funding and general support from the *National Geographic*. Later I followed Barbara's recommendation and read Mary Leakey's book *Disclosing the Past*, which, while not exclusively about her husband, is certainly revealing of his temperament and leanings. He was, first and foremost, a fund-raiser for the studies he favored: the excavation of promising anthropological and archaeological sites especially, but also the observation of lives and habits of man's closest relatives, the still-extant anthropoid apes. As an African himself Leakey, who had been initiated into the Kukuyu tribe when he was a boy, naturally understood chimpanzees and gorillas better than he did orangutans, but this did not interfere with his generalizing about the great apes, even the orangutan. Later, therefore, he encouraged and helped Biruté Galdikas to study orangs, and persuaded the *National Geographic* to give her a certain amount of support.

"Perhaps he also helped Dian Fossey and her mountain gorilla project," said Barbara, "but of that I am not sure. Dian is the only one I don't know about."

"But it was always young women," I commented.

"Yes. He probably had some theory that women are more patient and sympathetic than men," said Barbara. "Besides, he liked women. Well, he took me around the animal reserve near Nairobi to see lions and whatnot, and of course, I was flattered by his attention. He invited us to stay at the Leakey house and everything, but somehow—well, I found the man unattractive. There was no spark. He told me what to do in order to bring off the orang project in Borneo, and"—here her voice took on a tinge of indignation—"he translated the situation with gorillas and chimpanzees, which he knew from Africa, into the tropical forest of Southeast Asia, which he *didn't* know. It was a completely different situation."

Because of this, she felt very dubious about Leakey's judgment.

A sidelight on all this was that various young women who flocked to Leakey's lectures and were fired up with the desire to go out into the wild and live with apes and so on "came to consult me, too," said Barbara. "After all, I was *doing* it. I encouraged some of them. One of them especially, a young woman from Switzerland—she was determined, and she went to Sumatra and did it. This was after 1970, the year I tried to get back to the business of orangutans, but all that came later.

"Tom had to retire in 1967, so we went to Cornell, where he had a contract as a research associate in their Southeast Asia program." They went early so they would have time to look around for something else to do when this was finished.

"I enjoyed Cornell," said Barbara. "I attended lectures and got to know a completely new environment. Then . . . Tom had worked two and a half years off his contract when he suddenly threw it up and vanished. He had got to know another woman, a widow living in Belgium, who had independent means, and he simply went to live with her."

Barbara paused for a moment, then went on steadily. "He came back to

Cornell eventually to tell me about it, and that was all he did. It was a shattering experience for me. I had been married to him for twenty years."

Ultimately the Harrissons were divorced. "I told the department in Cornell what had happened—he had not done so—and the faculty was very decent. I had given lectures alongside Tom, and this arrangement had gone down rather well, so they said to me, 'We will give you the last of Tom's contract to work out.' There were six months left, and I got another year or two. At the same time people were saying, 'Look, you are good at this, but we cannot employ you permanently because you have no degree. So we will offer you a graduate scholarship for a Ph.D. and pay all fees and pay your cost of living if you can get through the course within six years.'

"Before I accepted and it was arranged, I made a move to go back to Sumatra and work for orangutans there, because after Tom left me, I had thought, 'OK, now I have time!' and I had a very good connection with the World Wildlife people; I had their assurance of Wildlife support for three years to build up an orangutan rehabilitation project on Sumatra. They paid my fare to go there and make all the arrangements. I went down there—and came back, having failed. That is because I was used to the uncorrupt government of Borneo, whereas the Indonesian government was very corrupt, and I found it impossible to make a commitment to the Wildlife people where in Sumatra I had to pay, right, left, and center. Not to mention their inefficiency and selling animals. I was too honest; I just couldn't make it. I had a row with the Wildlife chief out there, and after that row was finished, my future in Indonesia was impossible. I was spoiled, I guess, from my experience on the other side, in Borneo, with the British government that is no longer there now. Indonesia was quite different. I lacked the experience and the diplomatic touch to overcome that.

"So I went back to Cornell, where, as I said, I was offered a scholarship in anthropology with a background of apes and monkeys, and I was offered another in art history. I took the one in art history because I knew that in anthropology at Cornell you had to study more than three years. There was also so much work in theory in anthropology. It was interesting, but not something at my age to go in for for five or six years. Art history has always fascinated me; I had started it when just finishing school and never finished because of the war. I had very good museum experience, a lot in ceramics especially, so I had a very good specialization to work with. I did it by the skin of my teeth because all money was vanishing, and if I had had to pay fees, I couldn't have hoped to pay them. Tom was living in Brussels, and had sent me something like a hundred or two hundred dollars a month. Then he was suddenly killed, in a road accident in Thailand, alongside his wife, who was also killed. I was still in school at Cornell and not earning money. I wrote to the Sarawak government asking whether they could not make a special case for me in the matter of Tom's pension. While I was living in Sarawak, I had put in something like twelve years of unpaid work at the Sarawak Museum, so they might consider this as a special case. But they did not.

"I had to do something. I applied everywhere for jobs, and I got one in Perth, Western Australia, competing with something like sixty other people. It was very difficult to get in. I got in partly because at the time I was in Sumatra on that Wildlife situation, casting about for some type of job, and because it was not very far from Perth, I flew down there and gave a couple of lectures and made myself known. So I had the advantage of being a known person. So I got the job, for three or four years with the promise of getting tenure, but that promise was completely undone during my first year. There was a change of government; the buildup of universities and schools, which was the work of the socialist government, was stopped; and the student population was shrinking. Anyone not Australian had a very, very difficult time, so I knew I had to cast around for something new.

"This thing I have here is really an accident, due to my way of keeping friends. I have very old friends, and I keep in touch with them. This museum I have known since the fifties, when I was married to Tom and worked in ceramics. Tom introduced me to it. We had spent a holiday here in Europe, looking in museums and trying to find examples of big pots." Here she was talking about the huge pots found in Borneo and some other parts of Indonesia, in longhouses. "We came here in winter, and it was the one museum that impressed me very much. We went around and looked at the exhibits, and when we got back to the hotel, Tom said, 'You know, this museum is rather nice. The one thing I could think of to do after retiring when I am fifty-five is to take on this museum. I think this would be very nice to do.' I said, 'Yes, why not?' Since then I had come back two or three times, the last time just before going back to Cornell to take my finals. My predecessor was in the museum and I had written to him asking if I could come and look at the big pots, photographs and so on, and of course, he said yes.

140

"On the last day before taking the plane back to Ithaca, I had a talk with him over dinner. I was late bringing up a certain matter, because I had found that two pots the man had bought were no good. I didn't particularly like this man, but he was rather jolly, and he liked a bottle of wine or whatever we had; he was getting rather sozzled. So I started to tell him about the pots, and he, unlike some other people in his position, took it with great humor, and we embarked on a very interesting conversation. Just as we were breaking up, he said, 'You know, I have never talked with anyone as interesting as you about the pots in this museum. Why don't you become my succcessor?'

"I stared. 'Well,' I said, 'who is talking? Are you serious?'

" 'Yes,' he said, 'I'm retiring in two or three years' time, and that's an offer.'

"I went back to my hotel, and I couldn't sleep. I was completely bowled over. So I decided to cancel my flight and stay another day. I went back to that man, into his office, sober, in order to see if he remembered anything about it. I said, 'You know, we had a very nice conversation last night; I want to thank you for it'—because he paid for the drinks and everything— 'and you said something at the end that interested me. I wanted to know if you remember.'

"He said, 'Yes, I do remember. I was serious.'

"I said, 'I would be most interested in getting this job, but you must tell me when the moment comes, because I wouldn't know otherwise. I will tell you where I am, where I can be found, and I would be most grateful if you would let me know.'

"Two and a half years later I had his letter. I had never written to him because there was nothing to write about. His letter was very short, saying simply, 'The job is now open.'

"It was 1977. I struck my tents in Australia and moved here and never looked back."

I asked about her zoo visiting, and she said, "I don't like zoos, and I didn't like doing it; but I was in the International Union for the Conservation of Nature, part of the Survival Commission, a committee run by Peter Scott, who was an old friend of Tom's. In the committee I worked with a number of people who were concerned with international traffic in wild-caught and protected animals. CITES [she was referring to the Convention of International Traffic in Endangered Species] was not yet in existence; ten years before it was formed, I was part of a sort of club of these people.

Zoos were also represented on that committee, and they always followed the line that they were doing excellent conservation work because they were breeding rare animals, and so on and so forth, while at the same time they were playing a double game, buying wild-caught animals. You have another difficulty with zoo-bred animals: They have a very difficult future. No, it is something that does not attract me. What I did, essentially, was try to convince people that they must not buy illegally caught animals and must keep the ones they have in zoos in a reasonable way. It was a very different job from what I was doing in the forest, pioneering the project of putting the animals back. That is something which actually works and which continues working in Sumatra and Sambal. It was only discontinued in Sarawak because there was no wild population.

"I still keep up correspondence with workers in the field, you know. In fact, I went to see Biruté Galdikas when she was starting."

Here Barbara was speaking of the woman who might be described as the foremost worker today with orangutans.

"That was an interesting visit," she said reminiscently. "Biruté first got interested in such things after a lecture which Leakey gave at Cornell. Biruté was much stimulated by Leakey. I don't remember how she happened to attend the lecture; she was not a student, I'm pretty sure, but anyway, she heard him lecture. I think she had started studying at another university, something like that. Anyway, she was a Canadian immigrant— I think the name is Greek—and her husband, or perhaps I should now say her ex-husband, Rod, is Canadian. Anyway, she came to see me while I was still living in Ithaca and said, 'I'm fascinated by this study of apes in the forest, and I want to start doing it. I've talked to Louis Leakey many times, and he encouraged me. He said, "Yes, yes, come and do this project with orangutans," and he told me to do this and this, and you have to spend some years in the forest and get the animals accustomed to you; then you will have a beautiful time. . . .' So she was full of enthusiasm, and she wanted information from me how to get started; then she would do it. There was no doubt in her mind at all. I tried to explain to her that what Jane Goodall had done with chimpanzees was not so easy with orangutans because they are sitting up *there*."

Barbara pointed up to imaginary tree branches.

"Biruté paid no attention to that. She persuaded her husband, Rod, who was a nice guy, to come out to Indonesia with her and start this thing. So on my way to Australia or from it—I can't remember which, but I know I was doing something in Jakarta. She was starting with six months in Borneo before going to Indonesia, and I knew she was there, so I took a plane in Jakarta and went over to see her. There was no camp, no boat where they were; it was terribly difficult to get in. Rod came to walk me in, and he was a tower of strength because he was a very sensible man and had had experience in the forest; he knew all the plants. I don't remember what his background was. He was not a botanist, but he *was* a pathfinder of sorts. Walking, he said, 'We are in difficulties. Please try to do something for us.'

"They were living in a broken-down tiny little hut. Everything was leaking, and they didn't have a proper toilet. It was just beyond description, *and* they were living with three little orangutans, orphans that had been confiscated and handed over in the usual way; the orangs were tearing down everything that was not yet broken. Biruté didn't see any of this; I'm convinced that it was because of Rod she survived. He did everything; he got in the supplies of what they had of prepared foods, though mostly they were living on morsels and scraps. You know I can live very primitively. It doesn't disturb me for one moment, but they were doing it all the time; everything was falling down while Biruté was always disappearing and trying to do what Louis Leakey had told her to do.

"*And she was succeeding*," said Barbara impressively and unexpectedly. "She said to me every morning, 'Never mind, you go with Rod.' Then she would go off on her own. That was not a very attractive forest. Part of the time you were wading in water up to *here*. The leeches were not so bad; but it was extremely bad swamp, and to walk ten or twenty yards was a major effort. I had two or three days there. Rod always knew how to find Biruté, how to find the animals. Because it proved to be an attraction, she would take one of the little ones with her; I didn't advise her to do it, but it was a natural. By that time she was in touch with three or four wild orangs. She always knew how to find them. Essentially you had to see where they would make their nests at dusk and before dawn be at this place. Sometimes they would get out before dawn for some reason or other and shift in the night, so you lost them. Rod had found a way to negotiate the jungle at night and climb a tree in the middle of nowhere; he was very good at that. He had my sympathy, because he was more or less like me in the old days; he had helped Biruté by doing all the hard, boring things.

"I went back to see them once more later. By that time they were motorized and much better supported. I had succeeded in getting them support. Coming out of there the first time, I had been full of news, and I told the Wildlife officers, 'You must do something for these people; otherwise they will collapse.' This worked well, and I got the money. Now they had houses, they had guards, and one of the Indonesian students I had stimulated to go out, he was helping them. By that time Biruté had had a child by Rod, and Rod was getting fed up. Biruté told me that Rod had made up his mind to go back to his old studies. I don't know exactly what happened; but he went back, and I would be most surprised if he ever went back to Biruté."

She sighed.

"You said something about psychology outdating itself," I reminded her, "and then we were interrupted. It was while we were talking about primatology, wasn't it?"

"Oh, yes," said Barbara. "What I said was that a lot of these men and women are on a line of research that is outdating itself. They base their research on animals that are behaving completely in an unnatural environment, and they try to relate this to human behaviour, which is quite different biologically. So they create something that sounds interesting, without giving it the basis of continuity. That is how I see it."

I asked, "Why do these women do it? Not the psychologists necessarily, but the others—the Janes and Dians and Birutés? I know why *you* did. I know why I'm fascinated by their stories. But why do they?"

She said slowly and thoughtfully, "I think Jane, who started it all, got fascinated as she went along. Meeting Leakey at the time she did, it was inevitable. I went to see her back in early days, when she was married to that photographer, and I wasn't surprised that the marriage came to an end; after all, she was a much more important person than he. Dian Fossey I don't know about because I never met her, but she, too, was stimulated by Leakey in the first place, and then she went to see Jane, and it all followed naturally. Biruté, she met Leakey as I've said, and then she came and talked to me, and she saw Jane, and— You know, to go in and observe animals in the wild is something basically interesting to people. Many young people like the idea to start with. If you are very motivated in this direction, you want to go for the more difficult and interesting animals or animals that there is very little known about. You know, the chimpanzee bit. I guess people who become anthropologists are also motivated in this direction;

they want to observe everything. They are unsure of themselves. I have got to know quite a number of anthropologists and psychologists who are sort of half mad themselves. . . . I think it's a special makeup that leads one to study apes in the wild. Otherwise I think you can't explain it in a general way; you have to explain each case on its own merits.

"Yes, on its own merits. If Tom had not left, who knows?"

Then we looked at some of her small but choice collection of ceramics, and her nice new house, and her garden. Nowhere in the place was there any photograph or other memento of the orangutans.

9. *JO FRITZ: NO CHIMP IS AN ISLAND*

THERE ARE NOT A LOT OF PRIMATOLOGISTS or even primate fanciers in the United States, considering the size of the country, and word quickly gets around among them as to who might be interested in their rather special endeavors. No doubt that is why I began hearing from the Primate Foundation of Arizona; somebody put me on its mailing list. For some years I have been receiving in the mail its monthly newsletter, "Chimp Chatter," printed on pistachio green paper. The masthead describes the organization as "a non-profit, tax-exempt corporation devoted to the preservation, propagation and study of the chimpanzee."

The Primate Foundation not only attracted but puzzled me. Why were these animals in the middle of the Arizona desert? The foundation seemed to be without any university connections, although chimps and universities usually go together. For instance, Yerkes and his chimp colony, years ago at Orange Park, Florida, were backed by Yale, and later, when Yale withdrew its patronage, the colony was moved to Emory University in Georgia, where it remains as one of the National Institutes of Health's primate colonies under the old name. The famous sign language chimpanzee Washoe first got her training through the Gardners at the University of Nevada at Reno and later moved to a larger class connected with the University of Oklahoma at Norman. I thought of several more colonies all at universities. I knew there was one not far from Austin, where, of course, there is Texas's state university. There used to be at least one other exception at Holloman Air Force Base, but it seemed to me that I had heard it was being dispersed. Formerly there was a small number of chimps at Stanford, where Jane Goodall and another primatologist, D. A. Hamburg, attempted to reproduce a natural environment for four young chimps, but it came to an end when the funding ran out. Of course, I reminded myself,

Tempe, Arizona, part of the newsletter's address, is the seat of the University of Arizona. But distances are great in the West, and the foundation was not necessarily connected with the university.

By early 1978, through the newsletter, I learned more details of the situation. For example, the colony was nearer Mesa than Tempe. It has been named Malacandra, out of a book by C. S. Lewis, to describe a place of happiness. It is in an isolated area, "where visitors are not encouraged (even if they could find it)." (I quote from a column in the paper.) "You must travel almost 8 miles of dirt road and have detailed directions or a map. . . . Malacandra is approximately 40 acres of beautiful desert land, which is leased from the Federal Bureau of Reclamation. The desert is Lower Sonoran Desert (Saguaro Cactus, Barrel Cactus, Palo Verde and Mesquite Trees and Creosote Bushes). . . . The annual rainfall is approximately 9.01 inches. A red mountain rises straight up from the desert floor in Malacandra's front yard."

It appeared that the couple responsible for the Colony, Jo and Paul Fritz, often had more applicants than they could handle. In the issue for April 1978 were typical notices: "A 10-year-old male chimpanzee who presently resides at Marine World needs a new home. Contact [name and address]. . . . A 9-year-old male chimpanzee, retired performer, is still looking for a new home. Contact [name and address]. . . . Editor's note: The Foundation exists for chimps like these two, yet we were forced to reject them. Why? Lack of funds and lack of room. Help us to help them, please."

This place, I reflected, was something to investigate. I was aware that one of the greatest headaches in the animal entertainment world, which includes zoos as well as trained animal acts on their own, is what to do with mature chimps that have outgrown their cuteness and manageability. Chimps as they grow up get very strong and develop wills of their own. The experienced student of such matters knows just what it means when an experimenter in a scientific paper, talking of his work with such an ape, closes his report with the sad words, "Here the experiment ceased, owing to lack of co-operation on the part of the subject." In other words, the chimp bit the experimenter, trashed his equipment, or did something else horrendous, such as eating precious notes or breaking a window and running away.

Like many other wild animals, chimpanzees are getting scarce in their natural habitat. As their wild population shrinks, scientists must take better and better care of the captive animals; otherwise the world of research will find itself bereft of the most valuable creatures we have. Chimpanzees are so nearly human in physique, blood, and psychological reactions that they take the place of human beings (on whom, of course, it is forbidden to experiment) in many important laboratories. Yet every year laboratory animals grow older and harder to handle. In the old days, before we saw the world shrink, people were prodigal in their use of the apes. It seems only yesterday that performing and circus animals, if not the laboratory specimens, when they got too difficult to handle and too expensive to keep, were simply destroyed; the scientist or animal trainer just replaced his animal by buying another from a dealer who caught them in the wild. He can't do that today. Chimpanzees have at last been placed on the World Wildlife Fund's list of endangered species. The fund marches side by side with the International Union for Conservation of Nature and Natural Resources, and when its officials say that an animal is endangered, this means a good deal. For one thing, the animal cannot be imported into any country that belongs to the union; that means there is no profit to be had from capturing it. Also, being on the list means that chimpanzees already living in the United States automatically become far more valuable than they were, a fact which lent force to the conviction entertained by the founders of the Primate Foundation, Paul and Jo Fritz, that superannuated chimpanzees would be far better retired to breed than simply killed and thrown on the scrap heap.

Who were Paul and Jo Fritz? The story was told only intermittently in the monthly, much of the paper being given over to a feature that, as Jo Fritz later told me, is very popular but rather tiresome to write (she writes the whole thing). It is a letter purportedly from one of the chimp children, Geronimo, the foundation's first-bred infant. In the letter the latest excitements of the homestead, the birth of infants, temporary escape of the animals, and so forth, are reported as seen, one supposes, by a very savvy chimp, and signed GERONIMO in large, wobbly, printed characters.

"You wouldn't believe how many people on our mailing list want more of Geronimo," said Jo in a rather fatigued tone. I heard all this later, when curiosity impelled me to go out to Arizona and see it all for myself. She met me at the Phoenix airport, a tall, slim, fair woman who had warned me on the telephone to wear sneakers and rough clothing, and we drove, as she had also warned me, into the desert along a country road that entailed opening and closing a couple of gates. It was a rocky rather than a sandy desert, and some flowers were blooming on the cactus. The narrow red mountain was much in evidence for a long time before we arrived, standing out against a ribbon of blue mountains in the distance. As Jo drove, she told me the outline of how the place had come to be the Primate Foundation of Arizona, and we were not yet within sight of the two mobile homes that are living quarters and office room for the humans involved before we came upon a strong-looking edifice rooted in the pebbles of the desert floor.

"That's where the chimps live, most of them anyway," she said. "It used to be a hydroelectric plant. It's been adapted now."

As we drove in, a chorus of barks and hoots came faintly over the air. Jo remarked that they were being fairly quiet, as her husband, Paul, was over there cleaning out the cages and feeding the chimps.

"He's doing that all alone?"

"Probably not," she said. "My son, Jon, is helping out as long as he's here on vacation." Jon, she said, was a child by her former marriage, and he was twenty-four. "Usually some students come out from Tempe to help, too," she said. "I don't know what we'd do without them."

"Students from the state university," I said knowledgeably. No, said Jo, not necessarily. There are several colleges in the neighborhood, and the chimps, and Paul Fritz, seemed to exercise a fascination over them.

"The students are wonderful," she said enthusiastically. "Don't tell *me* they don't know how to work; they do work, awfully hard. But of course, Paul bears the brunt. He gets up at one o'clock every night to go out and feed the chimps, and it's wonderful how he can come back and go straight to sleep again. I couldn't ; once I've been waked up, that's it. But of course, it's his training. All his life he's taken care of animals, and I suppose it's second nature. When I'm raising a new baby whose mother has rejected it, as I am now, at first I have to give it two-hourly if not hourly feedings through the night, and it almost kills me. Fortunately it doesn't happen as a regular thing. I survive."

She laughed cheerfully and pulled up in front of the nearer mobile home.

"Here's where you sleep, in the office," she said, and we took my bag into the place and left it there, and continued talking as we drove over to the Fritzes' living quarters.

149

"No, don't try to walk on those stones till you get used to them," she said. When she pushed open the gate in the fence that surrounded the trailer, I saw a large burro nearby, and a small goat.

"We've got two African pygmy goats," said Jo, "and five dogs."

"I hear them," I said as three large dogs lolloped up to us, barking. But the noisier ones were two smaller dogs indoors.

Jo said, "When I told Paul I wanted a dog for Christmas, he hit the ceiling. 'Four dogs and you want another?' But on Christmas Day there she was; I'd seen her in a pet store, and I just had to have her."

"Does that donkey do any work?"

"No, she's too old, the poor darling. As for the goats—I'll tell you a story about my grandfather. He was one of the early settlers in Arizona, and I remember him as a very strict man. Well, he was, but I enjoyed staying with him and my grandmother on their farm. One day he came home with a young goat, and he said to me—he knew me pretty well—'Now, Jo, you're not to make a pet of that goat, because it's to be butchered for meat in the fall.' But I was only about eight, ten at the most, and I didn't pay any attention. I made a pet of the goat. When fall came and the butcher called by and got the goat, my grandfather made me go with him and watch it butchered."

I exclaimed, horrified.

"It was to teach me a lesson," said Jo. "So now I have two goats for pets."

She was very full of a fairly recent excitement that dated from February 15, when Arizona had been subjected to more rainfall in a short period of time than anyone there was accustomed to. Dawn that day, according to her account in "Chimp Chatter" for April–May, had been much like any other. I shall quote largely from the paper, as it told the story vividly.

> . . . except that the normally dry Salt River, upon whose bank Malacandra rests, was running with a moderate amount of water—and it was still raining. During the night "they" had started releasing water from the over-full reservoir up-stream from us. After three so-called "floods" we knew the amount now flowing was minimal and, in fact, for us was a lovely sight as it coursed through a desert lush from the many winter rains.
>
> At mid-morning, a friend called, who was very concerned. She said the radio had reported that the amount of water released would increase consider-ably. I reassured her, but as Paul drove in, from his daily round trip to Tempe, I rushed to him with the news. We were not terribly worried, but we did begin to monitor the radio and started what would become hourly

telephone calls to the agency that was controlling the water releases. By mid-afternoon, as the river kept rising, it was apparent that things might get a bit scary. I started telephoning volunteers to alert them to the possibility they would be needed. There was a feeling of controlled panic in the air. . . . By 4:40 P.M., we knew we would be forced to move the chimpanzees to higher ground before the night was over. As Paul said, "You don't wait to have the water up around your ankles before you start moving 30 chimpanzees." With only a tiny Datsun pickup truck and their own strong backs . . . Paul and Jon began to move everything portable that would hold a chimp even temporarily to the higher ground of the parking lot. The cages and boxes were not really large enough for one adult animal, yet we knew we would have to put two or more in each cage and shipping box, if we were to save them all. Paul made a list of compatible pairs and prayed. The last cage and box was [*sic*] in place by 8:00 P.M. and as the dark of night settled in, for what would seem an eternity, the first volunteers arrived. The river was roaring with a fury that increased steadily and the rain and wind were relentless. We might be saving the chimps from drowning, but how many would we lose to pneumonia?

A volunteer started the others building walls of straw bales around the cluster of cages and boxes and put over them sheets of plywood and any plastic cloth they could find to form a shelter. There was a new baby, Akimal, just five days old; he had been rejected by his mother, and one of the Fritzes' friends, Leanne Nash, an anthropologist from the university, took over the responsibility for his two-hourly feeding. So the women sat in the office, monitoring the radio, while the foundation's favorite vet, Jim Ebert, of the Health, Education and Welfare Center for Disease Control, with Paul and Monte, a volunteer, began to anesthetize the apes. Jo wrote:

Each had to be shot with the Cap-chur Gun and two must be shot in rapid succession so that as soon as they were anesthetized, they could be loaded into the back of the little truck, driven up the hill and placed together in the same cage. Dr. Ebert loaded the syringes with anesthetic, Paul selected and shot the chimps, Monte and Jim Baugh [another volunteer] watched and as soon as the chimp was asleep, they carried it to the waiting trucks (there were now two, thanks to a volunteer), where other helpers waited, shuttling them up to the cages.

Two hours later, when the last chimp had been carried out, several of the volunteers left. There didn't seem much left to do. Then the agency controlling the water release telephoned. Jo wrote:

[T]here was a risk that the released water might become less controlled and break over into the canal in front of Malacandra and upon whose banks we reach or leave the area. Malacandra could become an island . . . and the agency thought we should attempt to move out now. We couldn't—there was no point to even considering the suggestion. There were no trucks available at midnight and no way to load cages even if there were trucks. You must remember—these were full-grown chimpanzees in those cages and even with a team of men to lift—it was impossible. You would have to grasp the bars of the cages to lift and if you did, someone could lose a finger, hand or arm to a frightened, upset chimp. We could not move, but we still had the feeling that even as an island, we would be safe, the parking lot and office area were on fairly high ground. Yet—we couldn't seem to sit still and do nothing.

The anthropologist managed, after hours of telephoning, to secure from the chairman of the university anthropology department, a flatbed truck, which he himself drove over. Jim Ebert got permission from the director of the zoo to bring in whatever chimps they could and lodge them in the maintenance building. Five young people from the *Mesa Tribune* came to help load.

While the truck was being loaded, the governor of Arizona came onto the TV screen and ordered the evacuation of a very large area of Tempe and Mesa. There was deep concern, he said, that one of the dams that held back millions of gallons of water might have a structural defect. This meant that the Fritzes and their land would be inundated long before the rest of the area, for they were in a direct line from dam to flatlands.

At 2:00 A.M. seven young chimps, all that could be safely lifted, were loaded on the flatbed truck, and it left, bound for the zoo and followed by all but two other volunteers and Paul and Monte, who had to unload the chimps at the zoo and settle them in. Jim Ebert and Leanne Nash stayed with Jo and Jon. In tears Jo asked Jim to promise that in the last extremity he would put to sleep her burro and her two little goats. He said that he would and that if they all were forced to leave, he would take the three big watchdogs with him. Then Jo went to the house to fill a grocery bag with whatever seemed important. She did her best but couldn't really think clearly. What, after all, was a treasure and what wasn't? She went back to the office and with Leanne's help began packing important records of the foundation. It was estimated that there would be a six-hour notice downriver if the dam went, which, they figured, meant two hours for them.

At 5:00 A.M. Paul and Monte came back from the zoo to report all well.

They were exhausted, but as the sun rose, everybody felt better. The rain stopped. Paul and Jon went off to rent two more vans. The three watchdogs were loaded into Jim's van, and he went home at 6:00 A.M. because his family was worried for his safety. The chimps, especially the babies in the nursery, were waking up and making hungry noises. It must have added to the general sense of unreality, I thought, when a television helicopter set down its crew to see how they were getting on and ask when they were leaving. The men took some photographs of the nursery babies, then took off again. Eventually Paul and Jon returned with a rented van and said they had another waiting. They parked it and went back in their little truck to fetch the other one. Another vet arrived with a horse trailer and fetched the burro and goats, so that much at least was off Jo's mind. Little by little vans were assembled, big chimps were loaded without incident, and all was ready. After their first food in twenty-four hours the Fritzes, too, were ready to go. All the animals were accounted for except a coyote Jo had been taking care of for the fish and game department people, who had picked it up as a small cub and asked her to look after it until he—it was a male coyote—was old enough to shift for himself in the wild. They had left him far too long, and when at last he was supposed to get out into the wild ("Jo," Paul had said warningly, "I don't know any other man who lives with a coyote under the toilet"), he had refused to leave. To save him from the dogs, they had built him a run, and there he lived, stubbornly refusing to take advantage of all the chances the Fritzes gave him for freedom. But now, Jo was happy to see, he left the run and really went into freedom on his own.

By 7:00 P.M. everybody had left Malacandra.

They stayed away six days. The dam held, and then they came back. "The move cost the Foundation over $2,000," wrote Jo, "but we'll manage somehow—it isn't important. What is important is that not one chimp even so much as caught a cold while living in those vans and our two pregnant mothers are still pregnant." All the other animals were returned safe and sound. Even the coyote put in an appearance, ate from Jo's hand, and silently departed on his own affairs.

"He returns every ten days or so," wrote Jo, "just to say hello and to assure me that he is alright [sic] and then disappears again."

Before Paul arrived, Jo began to tell me more about him. He was an animal trainer, born in Germany, and she met him when he came to the

153

Phoenix Zoo to work a little animal show that he had been traveling with. Jo was a volunteer worker in the animal nursery; in fact, as Paul later told me, she was the head of the volunteers there.

"I did his announcing once he'd set up the show," said Jo, "you know, for the chimps' tea party and things like that. We had a tiny baby elephant somebody donated to the zoo, and the three chimps Paul brought with him, and a young leopard, and so forth. Oh, that elephant was sweet. We put on a charming little performance for a season, but I'll wait and let Paul tell you about himself. Let's see how the babies are doing."

Three young chimps kept each other company in a playpen in the Fritzes' sitting room. All three wore diapers, but one male, Babad, was also dressed in a romper suit.

"We have to keep that on him because that bigger female, Hahshani, strips him otherwise," explained Jo. "The third one is Jayme, and I wrote about her in 'Chimp Chatter'—the little blind one, remember?"

I did remember. Jayme, for no reason anyone could understand, was born blind. An eye doctor, Edward L. Shaw, who lives in Phoenix, heard about her and offered to operate on her eyes free. He did so and was successful in restoring sight at least to one eye. I looked at Jayme, who was obviously low man on the totem pole; the other chimps walked over her as if she weren't there. Jo said, "After all, she's a year old—tomorrow is her birthday, I think, or is it the day after?—and she was blind a long time. She got used to it, and it's hard for her to be used to seeing now. She does see. She can play finger games with me, and her eyes follow colors. But she's accustomed to getting around the way she used to, by feel and smell, and we just have to keep working on her so she'll learn to be a chimp. Won't we, Jayme? Come on out for a little and lie on the rug. I know she recognizes me by smell and touch, because she comes to me right away if I'm not wearing perfume. If I am, forget it. We just have to keep on trying to stimulate her to use that eye. It takes patience."

"There's a portrait in that room where I'm going to sleep," I said. "That benevolent-looking old Gladstonian gentleman with gray chin whiskers—who is it?"

"Danny," said Jo, and sighed. "Our wonderful Danny. I really think the foundation wouldn't be here at all if it weren't for him. He was so wise and so gentle—the real old man of the family. I often asked his advice; I depended on him. If a chimp was a newcomer and was upset or something,

Danny comforted it by patting—pat, pat, pat— until it quieted down. But he's not with us anymore. He died of a disease we have here called valley fever. And—wouldn't you know it?—there's a medicine for it now, though it's not on the market yet for animals."

In the kitchen she scalded bottles and prepared four different formulas. Three of the bottles went to the three juveniles in the playpen, and then she went into a back room and emerged with a very tiny chimpanzee that was rather listless about taking its bottle.

"Could I try?" I begged. "And what's his name?"

She handed him over, and after fussing a bit, he settled down. "His name's Aki, short for Akimel. All our chimps have Indian names; we let the Indian children name them," said Jo. "Akimel's the Pima—or is it Papago?—for 'river.' Babad means 'frog.' Hahshani means 'saguaro cactus.' "

We sat comfortably at the table drinking coffee, all except Aki, who was by now taking his milk. "He was never his mother's first interest," said Jo. "She handed him over to Paul as soon as he was born, and Paul handed him straight back, saying, 'Here, I haven't got time. Take care of your own baby.' But she wouldn't, and then we had the flood and everything, and that settled it. He's been on my hands since."

"Ah, already a mother!" said Paul Fritz to me as he entered the room. He is a very wiry man with what is called a sensitive face—that is, he is so thin that his cheeks are hollow. But obviously he is very healthy. He sat down and drank coffee, and after I asked him to tell me about himself, he started somewhat shyly to do so. He was born in Berlin, across the street from the famous city zoo. It was inevitable that he should be interested in animals and, as he grew into adolescence, should seek work at the Gardens, as the zoo was called. He learned to know all of the animals, though at his tender age he could not aspire to be a full-fledged keeper; keepers in Germany before the war, he said, were a very special lot, and their training was rigorous. The war came while Paul was still too young to be drafted, and at any rate, the authorities were proud of the Gardens and did not strip them of manpower until all else was forgotten in the final days, when Paul was at last drafted and had a few days of service. Later he drifted about, getting jobs with animals wherever he could. This led, inevitably, to work in circuses. He had the chance of a good job with the English animal park and circus combine Chipperfield, until it turned out that he couldn't go to

England because no Germans could get work permits there. Nevertheless, Chipperfield's was anxious to secure his services and got him to go to France, where by way of a French passport he might at last get into England, but that fell through. Ultimately he was permitted, like other Germans, to work in England, but that came later. In the meantime, he took what jobs he could get.

"I never had the hope to work in a circus," he said. "For a time I was at a zoo where I lived in the monkey house. Of course, there were no monkeys; they were all dead, and that is where I lived. The black market was going good, which is how most people managed, by swapping. For instance, I had a lot of electric light bulbs, and I exchanged them for things I needed. One day I went into the country to a circus with some bulbs to exchange them for food, and I found a woman there in charge of a circus, or what was left of it. Her husband had died, and she needed help. There was rationing for animals, though not for us, and there was one lion at the circus, the only survivor of all her lions. Those who worked with the animals shared their rations; thanks to that lion, I got more meat than anybody outside had had for months. So I stayed and helped the woman, and she helped me. So that is how I got into circus work."

In the course of his career, Paul said, he had found it necessary to work with all kinds of animals, so by the time he got to Chipperfield at last he was able to manage a large array of charges to train and put onstage.

"There were three polar bears, eight black bears, eight lions in another act, four tigers in another, a dog and pony act. . . ."

I quoted the famous circus animal trainer Günther Gebel-Williams, who once said that bears are hard to train because one doesn't know from their impassive faces what they are thinking. With all due respect to Gebel-Williams, said Paul, who is an old acquaintance and thinks Gebel-Williams the very best trainer ever, one *does* know. Gebel-Williams himself knows, because he is a trainer. "What one does," said Paul, "is think like the animal. If you are working with an elephant, you get into that elephant's mind; you think like him. If you are working a lion, you think like the lion. You have to switch over and become that lion. I agree it is harder to work with bears, but I like it. I once owned two small ones; I think they are very high in intelligence."

"But you say that about all animals," put in Jo.

"You can teach a bear all you can teach a chimpanzee," said Paul firmly.

In due course Ringling Brothers came and hired him away from Chipperfield, and thus began his wandering all over the circus world. "I met Günther at this time," said Paul. "He is wonderful, truly. I have seen him after a performance, and he is just as keen then, just as interested in his acts. He is a real professional. He is marvelous."

I mentioned a polar bear act I had recently seen at Ringling's, a number of huge animals presided over by a girl. He nodded. "Yes, I know the act, and a polar bear is the most dangerous animal there is. I learned about training them from Hagenbeck, the old man of that circus. He was seventy years old, but he certainly knew his business. He told me they got their bears from Norway, from the whalers who used to catch baby polar bears when they were swimming near the whales. Those little bears were running around in their cage at Hagenbeck's really wild, ready to eat you if they could, but the old man used to take care of them and train them. He had seventy-four of them at once in one cage. He always worked in tails and white gloves, and he would take off one glove and wave it for his signals. That's the only thing he used to direct those bears—one white glove. No doubt you saw the little arena at Hagenbeck's for shows; it's really just an animal school. The good trainers worked in the ring. Nowadays you have animal behaviorists who get Ph.D.'s on studies of animals, but they never come close to them like the old-time trainers. I was lucky to have some of that experience very young. The Berlin Zoo had a big outdoor lion act, you know."

Ringling's underwent a great change in 1966, as circus buffs will remember, when it first performed under the roof of Madison Square Garden—at least one of its circuses did. The whole picture altered from then on. Whether because of this or wanderlust, Paul moved on to a circus in Havana, and later he accepted a job in Australia because he was curious to see the place. He certainly saw it. The circus went everywhere, but unlike Ringling's, it depended for motive power on its five Asian elephants, all of which were, of course, under his direct care.

"The big top," said Paul, "and one tractor, to pull the whole thing between engagements; everything else the elephants did. They loaded the train, unloaded it at three o'clock in the morning, carried the stuff to the circus grounds, put up the tent, performed, carried everything back to the station afterwards, and loaded the train. . . . And this went on and on. My helpers were whatever I could get on the place, alcoholics, lunatics, anybody; I never dared to go around the circus grounds without a big stick in my hand. I got stuck with those elephants, I admit it. I like elephants, but not in those circumstances. The circus was owned by five sisters, elderly ladies who never agreed on any point and kept changing each other's orders. After thirteen months I broke my contract, which was for two years, and came back to America." He laughed. "Back from Australia, and I didn't know what to do next."

Here he broke off to give Jo a hand with the formula bottles for the nursery chimpanzees. When he came back, he resumed. Somewhere along the way, he admitted, he had not mentioned his experience—and heartily favorable opinion—of the Bronx Zoo, but never mind, he had worked there for a while. Now, back in the United States, he settled down with his first wife in Sarasota, Florida, where circuses have wintered since American circuses began.

"For a while I wasn't working, just living," he explained. "I didn't know exactly what I wanted to do. I did know I didn't want to work in a circus anymore, or a zoo either. Well, there was a museum of animals not far away from our house, and I went and convinced the people running it that they should have live animals, too, just a few. I persuaded them to get two baby chimpanzees for me, and I trained them. My first experience with chimpanzees had been with the Berlin Zoo, of course. Well, they bought them for me, and I bought another for myself, exchanging four sulfur-crested cockatoos for it. I trained them, and then I had a chimpanzee act. I bought the first two from the museum, then traveled for ten years with my act, at fairs, nightclubs, circuses, with my chimpanzees." Along the way somewhere he and his wife got divorced.

When he reached Phoenix and was offered a permanent job there with the local zoo (with his apes, naturally), he accepted. The operation was enlarged by lions, some birds, and a baby elephant (as Jo had said) plus whatever else seemed useful. That was when he met Jo, who was running the zoo nursery; he annexed her to be the announcer for the act. It was a really nice show, he said. Of course, he still had his contacts in the traveling show crowd, and one day an old friend, a Hungarian who traveled with performing chimps, telephoned him for help. This was in 1970. The Hungarian wanted to retire, but he didn't know what to do with his three performing chimpanzees, females of varying ages. Would the zoo buy them? No, said Paul, the zoo would not buy them; it had no facilities for more chimps. Then would Paul, or anyone else of his acquaintance, be willing to take them as a gift? Otherwise the Hungarian, at his wit's end, would have to put them to sleep, as he expressed it, and it would break his heart. Paul was in a quandary.

"It was very hard," he confessed, "to think of those chimps with nowhere to go, and I'd known the old man for a long time."

Then he thought of what was at best a temporary solution, though how it could be finally settled he had no idea. He lived in a small apartment house downtown in Phoenix. It was owned by an animal lover who lived in the house himself, in another apartment. The lobby had never been adapted for living space, and Paul got permission to keep the chimps there, at least temporarily, and he moved the animals, cages and all, into the space. Nobody realized it, but this action was the beginning of the Primate Foundation. Jo helped him take care of the chimps, but it was not easy in those surroundings.

"It was because those first chimps were performers and belonged to an old-time friend," said Paul, as if he were still wondering just how it all came about. "Then we had a call from another zoo that had heard about us, and they sent us Dolly. Soon there was another call, from Hawaii this time— the director, Jack Throp, used to be director of the Phoenix Zoo, a very good man—and he wanted to get rid of his chimpanzees because he had to make room for gorillas. He had three, a male and a female and a baby. The male's name was Lancelot Kanaka, representing the best of England and Hawaii, Jack said. So we had all these animals—"

"Still in the lobby?"

"Still in the lobby. We had then five, six, seven chimpanzees. Something had to be done to get them out of there. It was quite impossible the way things were, though Jo and I worked very hard."

Forced to do something, the two chimp guardians spent all their leisure time, which was not much, looking around for roomier accommodations. They now had an idea: the ready enthusiasm with which their world received the idea of giving them chimps showed them that there was a crying need for a hostel for the animals. It should be a reservoir into which deserving researchers, behaviorists and the like, might dip. (Naturally Jo and Paul, then as now, were bitterly opposed to destructive experimentation.) Out of his rich experience Paul could see well ahead of most that the chimpanzee was rapidly becoming rare in the wilds of Africa, no matter how many went begging for space in America. They were already on the so-called threatened species list and would soon, inevitably, be moved to the next and last class, the endangered species.

Jo had a brother-in-law with a chicken farm at Tempe and nobody to run it. If Paul was willing to look after his twenty thousand chickens, he said, the chimps could live on the same land until a better place was found for them. So that is how it was settled for a time. The chimps, with Jo and Paul, went out to the chicken farm, and when Paul was not feeding chickens or running his show, he and Jo drove around the desert looking for permanent quarters. They found it finally in the hydroelectric plant near Mesa and arranged a lease, by no means swiftly, with the authorities. Along the way, just before they moved, they found the time to get married.

Then they set to work to find the money necessary for remodeling. They also started working at it. The first grant they ever got, on a dollar-for-dollar basis, stepped up their efforts if anything could. Necessary alterations were made in the old stone shell of the plant, and the chimps were moved in. There it was—a primate foundation. At that time, unbelievably, the number of their animals had swollen to twenty-six. They have far more now, but a lot of them are usually away at the laboratory known as LEMSIP in New York State, where they are used by serologists to discover the causes of hepatitis.

"We had to look a long time for LEMSIP," Jo told me. "No ordinary laboratory would consent to meet our conditions. We've got to approve the work, decide how long they can keep each chimp, and call them back whenever it seems indicated. LEMSIP agreed to everything, and it works out well. They really do take care of the animals. Just now everybody in the medical world, or almost everybody, is interested in non-A non-B hepatitis, and our animals are very important to that research."

Perhaps as a respite from talking, Paul took me out to see the chimpanzees' home, to the hysterical hooting of the big chimps when they saw him coming. We went in and stood in a complicated kind of corridor between a lot of interlocking cages, and one enormous animal swung back and forth on his feet for our benefit while we watched admiringly. On each swing he slammed his walls to a definite, well-kept rhythm, looking anxiously toward us as he danced, waiting for applause. Obligingly we clapped.

"He used to be a performer," said Paul. "You can see, some of them are able to move around with others when they feel like visiting, and if they want to be alone, that, too, can be arranged. The bars slide up and down when I want to let down the gate, but chimps hardly ever want to be alone unless they don't feel well. Here is where we keep the feed."

He led the way to outside rooms crammed full of crates of vegetables and fruit. Paul explained that he had a regular route among the groceries and supermarkets of the town, where they were glad to have somebody who would cart away their discarded produce. Chimps don't care at all when their mangos, avocados, bananas, and apples look spotty, but people do. The vegetables and fruit in the chimp diet are augmented by a prepared monkey food, and fibrous food is included every day. Cabbages are best for that, said Paul, because celery tends to stop up the drains.

"Chimps love yogurt," he said. "They're crazy for it, especially apricot yogurt, but it's got to be some kind of fruit flavor, like strawberry, loganberry, like that. They don't like it plain."

The apes' health is usually very good, better than it could be in a zoo, because the place is not open to the public, which brings cold germs and such with it.

We returned to the chimps, and Paul, pointing to one of the animals, said, "There is Geronimo, our first-bred baby."

"He's not a baby anymore," I commented idly. Paul pointed out the two expectant mothers, and I hoped they both would prove to be good, responsible parents, for Jo's sake.

When I got back to the foundation office, I looked up a newsletter I remembered from November 1976, in which Jo had listed the animals then resident at the foundation. I found their histories interesting. Part of the register read:

Name	Sex	Age	Background
Anne	F	17	Retired Circus Performer
Bambam I	M	6	Phoenix Zoo Performer
Bambam II	M	4	Houston Zoological Gardens
Betty	F	18	Houston Zoological Gardens
Carla	F	11	Los Angeles Zoo
Chiquita	F	15	Retired Circus Performer
ChipChip	M	25	Woodland Park Zoo

and so on.

"It wasn't easy, finding this plant," said Paul as we picked our way across the water-eroded stones that make up the valley floor, back to the mobile home. "We traveled miles and took a lot of pictures of possible sites. In fact, I didn't recognize the possibilities of this place till we developed the pictures, and then I saw how perfect it could be. It took just about a year to get the necessary permissions and all that. Added to which, the apartment house owner decided to get into the act on his own account. He liked having chimps in his house, and he's a great show-off; he wanted to keep on showing them to his girls, so he started a chimp house of his own across the valley, or tried to. He has money, so there was no difficulty there, and he started building. We asked our attorney if there was anything we could do, and he said, 'Get your chimps out there first,' so we had to hurry."

"We put up the cages in the building very fast," said Jo, pouring out coffee at the table.

Paul said, "We put up the cages before we got the lease settled, in fact."

"Every night, after Paul had finished with the zoo, we came over to work with the chimps," continued Jo. "There was a shack on the land at that time, and a man was living in it. One night he came down to the plant with a gun. He pointed it at us sort of waveringly and said, 'I don't want you, and I don't want your monkeys. I'm gonna blow you right out of your saddle.' And I said, 'He isn't even *in* a saddle!'" The Fritzes laughed. "So Paul, at whom he was really pointing the gun, told me to run and get the police; he wanted to get me out of there quick. I ran out to the man's truck, got into it, and drove up to his house. There was a cat in the truck and a bottle of vodka. In the house were bottles and bottles of vodka, but nothing to drink with it—no mixer, no bottle opener, no nothing. But there was a telephone, and I got through to the police, and they finally came and talked the man out of there, and my brother-in-law came, too— he's moved to Illinois now—and between them they calmed the man down but neglected to take his gun, and a couple of nights later he came back, waving it around just like before. Still, we got things done, and so on the night of Labor Day, after Paul's last show, we went in. We had police escorts just in case, and we'd borrowed their Cap-chur Gun from the zoo, and we tranquilized the chimps and started to move them in. There were volunteers to help, and several trucks. Every time a chimp would begin to nod off, some nice policeman would come by and shine a flashlight into the cage and say, 'Everything all right there?' and of course, the chimp's eyes would fly open. . . . But they meant well, and finally we got all the chimps into their rooms. And—here we are."

It was hard going. It still is, but there are now mitigating circumstances. The Fritzes had to pay all expenses out of their own pockets—chimps' food, upkeep of the truck, cost of the lease, everything—until they got their first grant. Most people who brought in their chimps didn't help with money. They still don't. They would come to the door, leave their chimps, and, in Paul's words, "kiss them and forget them." One animal donor did pay twenty-five dollars down when she brought in her discarded pet, but since then there has not been a word out of her, not so much as an inquiry. It is not quite like putting your elderly mother into a retirement home that requires all her money down in a lump sum; it is not at all like that.

"People come in all tears," said Jo, "saying they don't know how they can bear parting with their darlings. I wait for a day or so and then telephone to say that the chimp is settling down nicely. After that—complete silence. You would think that some of them, at least, could afford a little money every month, wouldn't you? But very few send even that. If it hadn't been for the subscriptions from people out of the blue, I don't know how we could have survived those first months. Little sums, most of them, but they came in awfully handy, and then there were the clubs and associations interested in primates. Besides, if we made people pay to put their animals under our care, they just wouldn't; they would get rid of them instead, which is what we didn't want—killing those chimps. One of our first lot of animals came from a laboratory that would otherwise have killed off the apes, and we had to pay the airfreight to prevent that, for ten chimps, can you imagine? The very first grant proposal came from a small local foundation. I had to write it all out. I'd never written for a grant, and I hadn't any idea what to say. We had to put in electricity, so I said that, and drill a well, so I put that in, all of it, and they came through with a grant of eleven thousand dollars!"

("Because Jo got that grant for eleven thousand," Paul told me later, "I rewarded her. I let her buy that burro for twenty-five dollars.")

"I couldn't believe we'd been successful, even after I worked so hard on the application," said Jo. "I nearly went crazy with joy. I called Paul at the zoo to tell him, but they couldn't reach him, and I called my mother, but she wasn't home. Still, the news got around fast enough."

"So then we could start our important building on the premises," said

Paul, "with volunteer labor and secondhand materials, whatever we could get."

"I had no idea how much things cost," Jo said. "I would say, 'Oh, five hundred dollars ought to cover that,' and it turned out to be two thousand. The drilling alone was horrific. . . ."

"Without the students it couldn't have been done," said Paul flatly. "What we built was good, I admit, but it could be better. When we have the money, we'll do it better."

"The students are so full of life and so opinionated. They love to sit on the floor and talk things over," Jo said. "It's fun to watch them grow up. They keep in touch with us, too, after they've gone away."

I asked, "Are they all in anthropology?"

"Goodness, no, most of them aren't. One was in a journalism course but had dropped out for a couple of years. Now he's getting his law degree," said Jo, whose son, Jon, was leaving next day for college.

Paul told me of one youth who went up to work on the Alaska pipeline in the summer and telephoned now and then to keep himself informed as to the chimp colony. Another is now at the Mayo Clinic and recently telephoned to tell Paul all about his work.

Paul added, "We have a saying in Germany, 'As I get to know people, the more I like animals.' Something like that. And it is true, but these students are good. I begin to like people, too."

It was time to feed the baby Aki again. He was teething; that probably accounted for his fretfulness and reluctance to eat.

"If your mother would only look after you, we wouldn't have this trouble," scolded Jo. Aki was returned to the nursery, where he had a special cot made out of a carton, and Paul went to snatch his midday nap. Jo told me that she was expecting company next morning: her favorite vet, Jim Ebert, and a man from the National Institutes of Health (NIH) in Bethesda, Maryland. He was David K. Johnson, D.V.M., who just wanted to talk things over for a bit. She had recently received the news that she was to be a part of the Interagency Primate Steering Committee, an ad hoc body, she explained, which was meeting in June under the auspices of the NIH. This was to be the second meeting; the first had taken place in 1978. Jo was to chair it, and she was nervous. She showed me the report on the first meeting, entitled "The Task Force on the Use of and Need for Chimpanzees of the Interagency Primate Steering Committee." It had been, and the new meeting would probably be, sponsored by the Blood Bank at the NIH Clinical Center, the Food and Drug Administration's Bureau of Biologics, the Center for Disease Control, two divisions of Research Services, the National Heart, Lung, and Blood Institute, the National Institute on Aging, the National Institute of Allergy and Infectious Diseases, the National Institute of Mental Health, and so on. The publication of the committee's *National Primate Plan,* produced in October 1978, provided in its contents a succinct statement of the status quo for that time:

STATEMENT OF THE PROBLEM.

    A. The Continuation of Many Essential Health Activities Depends Upon the Use of Nonhuman Primates.

    B. There Now Exists a Severe and Long-Term Shortage of Primates.

    C. The Primate Shortage Has Not Been Met by Private Enterprise.

    D. Federal Action Is Required to Deal with the Problem.

"Do you think we're hopelessly inconsistent?" asked Jo suddenly. "It's a very touchy point with some people. The so-called animal lovers who can see only the sentimental side of conservation—"

"Vegetarians who wear leather shoes and carry leather briefcases or handbags," I said, nodding.

"That's it. It's better not to get into arguments with them, but I have done it in my time. I say, 'Look, all you know about medicine, and diet, and the proper care of the young—that didn't come down from heaven on a cloud, you know.'" She brooded a little. "Take the IUD, for instance, the intrauterine device. They've been trying them out on chimps, and some of the things that happen to those animals would make your hair curl. Of course, it happens to women, too. They're working on something less lethal to use than the IUD, and they'd better come up with it, or else."

"I think they're giving up metal things," I said. "They use plastic, which seems to be better."

"Research. That's a fighting word with some people," said Jo.

"A dirty word," I agreed. I told her about an undergraduate girl who, during the protests of the sixties at Yale, had been apprehended in one of the biology labs at night, tiptoeing along with a big hammer in her hand. Caught, she explained that she had meant to destroy the plumbing in a room where an important experiment was going on to keep certain organisms alive in running salt water. "It's bad," she had said in explanation. "It's science."

"Years of work on that experiment could have been wasted," the professor told me. "Is the world going crazy?"

Next morning the two veterinary surgeons arrived and had a long talk with the Fritzes. My attention was drawn to Jim Ebert's hand, which was badly deformed, fingers piled up on each other.

"Good Lord," I said, "what did that?"

Jim said that it was done by a chimpanzee named Jake and that he was soon to have an operation that would correct some of the worst damage. Later I asked Jo for more details. She said, "Jake is a big male from Holloman Base, and he's interesting because he has what's called the Australian antibody against hepatitis. He was in his cage at Jim's lab. They had a young female on the floor, and they were preparing to lift her to put her into her cage. Jake began to display. Jim walked over to talk to him and tell him it was all right, and he laid his arm along the top of the cage, and Jake took hold of the arm and pulled it into the cage and twisted it and bit Jim's hand. Jim understands now, because Paul explained to him exactly what was happening. Jake was trying to protect the female, not because they were going to hurt it but because it was *his* female. What happened was that Jim had walked over at the beginning of the display, and at the end of it Jake, instead of pounding the cage as the final burst of aggression, just grabbed."

"These chimp display tempers are extraordinary, aren't they?" I said. "They just go blind."

She said, "We can always tell when one of the males is working up to it. They start to breathe hard, uh-huh, uh-huh"—with increasing force— "and then WA-AH-AH! At this point, when we try to stop it—we've done it to see if we could—we can't. It's like a train has started."

"Do you think it's what we call seeing red?"

Jo hesitated, thinking hard. "I don't think it's like that. It's a dominance thing. I've asked Paul the same question. Sometimes they're doing it over nothing, really nothing. Even when it's over some very small thing (in my opinion; of course, they have an entirely different idea of what's important and what isn't), it would seem, if they're really as intelligent as many people give them, that if you could say, 'Oh, wait now, *just* a *minute*,' they should be able to stop. But they can't. I wonder if they realize that you want them to stop. There are some things about the chimp that are a complete and total puzzle to me, because they are *so* intelligent, and yet there seems to be a block, areas into which it seems not reasonable not to carry over. Perhaps they're not so intelligent as we think. Basically, of course, they live by instincts that we've lost. Chimps are still a puzzle in many ways. And they are better than we are at knowing how *we* are going to react in a given situation."

"Is it only males that display?"

"No," said Jo, "we have females that display, too, but it's different. There's more screaming, for one thing. I have an analogy: that the male is like a big, burly, loudmouthed truck driver who is really, underneath, a softy. Describe the worst traits of the human male—the blowhard, the egotist—and that best describes the male chimp. Describe the worst in females, the things that people say are characteristic of women—they're sneaky, they're backbiters, they're gossipy, they're hysterical—and there you have the female chimp. It's very unfair to say that, but it makes clear the difference between chimp sexes. We have female troublemakers, like the chimp that goes to the dominant female and screams and wrings her hands and points to a chimp in the corner who's absolutely innocent, and the dominant female goes and beats up on that poor chimp who is sitting there innocent as innocent can be. And it's just because the troublemaker wanted the innocent chimp beat up on and didn't want to go and do it herself." There was genuine indignation in Jo's voice. "I have to use human words for it; you'll make allowances for me. . . . So you do have troublemakers, you do have liars. She's gone and told an untrue story. And you see it happen. You can see why. The hierarchy is established, and she wants, in her middle position, to prove she's the best friend of this one. You see children doing the same thing. '*She's* my best friend, and *she's* my best friend, and I don't want them to get along together or I'll be out of it.' That's why three is always a bad number for children and for chimps. There's always an odd man out."

"What's a good number for chimps?"

"Oh, it depends on the chimps. Some people say males can't get along, but they can, provided they're picked by Paul. In one room we have ten, and they get along fine. In another we have four, and we cannot integrate any other chimp in that group; we've tried, and it never works. What you have to do then is disband the group and place them, one by one, in other groups. Groups aren't inflexible in nature; they shift and change. Let's ask Paul about all this when he comes in."

In the meantime, we discussed something else that he had said when the vets were there. David Johnson had announced that there are more chimpanzees than anyone might expect in Sierra Leone. The question before the committee, he said, was what to do about it. Should they invoke one of those laws that punish people for killing and eating members of endangered species, or what?

"Because there really are too many chimps for the good of the countryside," said David.

Jo thought about it for a minute and replied, "In my opinion, no. Those people in Sierra Leone need protein. Who are we to tell them they can't kill their own animals when there are too many of them?"

I had been somewhat surprised at the time by her reaction, and now we talked it over. She was willing to stand by her argument, said Jo; the Endangered Species Act surely was never intended to victimize people whose country's fauna was numerous enough not to be called endangered. After all, in the last analysis people are people. . . .

When I could catch and corner Paul, I asked him my questions: "What goes on in a male chimpanzee's display? And can it be arrested in full flood?"

Paul said thoughtfully, "Well, if he is displaying to a female, perhaps it can be stopped. He is really flexing his muscles, bluffing; it's a lot of bluff, a display. Like a person with a lot of feeling, who's got to go through what he is saying and showing. . . ." We agreed that it could be called a rush of adrenaline. "Usually it can be stopped, but it depends by whom," he continued. "One male can stop another. It begins with the heavy breathing, but they use everything—every ounce of adrenaline, every bit of strength. It is very important to them. It decides their place in the hierarchy. They want to make a great impression. But they don't always get away with it; they show how great they are, but they don't always get their way. It's just a lot of bluffing. I *have* stopped a chimp's display. It's no good saying 'There, there,' until they have simmered down; it's a kind of confrontation instead. You say, 'I'm stronger than you are,' and then he sits down. Sometimes he starts crying."

"But when you're training?" asked Jo.

"Oh, yes, of course, when I'm training, I've got to stop it. I *have* stopped chimps in certain circumstances, but it must be on his territory. Otherwise I am as much behind bars to him as he is behind bars to me. I cannot threaten him if I am outside the cage."

"I see."

"But no rage is involved?" asked Jo.

"That uncontrolled rage when you are getting blind with anger? I don't think it is," said Paul. "The chimp is trying to be better than any other one with screaming, with showing his strength. I don't think it is rage. It looks like it, it is supposed to be, but it isn't."

"A kind of ritual?" I suggested. "Jane Goodall says that in nature the attack begins after the chimp has plucked a leaf and put it between his lips—"

"Or he grabs a twig or something else that he can throw. Anything that makes him look bigger," said Paul.

"We've seen people who put on a show like that," said Jo. "A person can start out rather calm, but he works himself up."

"Female chimps get hysterical maybe," said Paul. "They throw themselves on the ground, roll around, gasp. You see children do it, too, when they're in a temper. But the male does not really get hysterical. He's just a blowhard."

Next morning Jo and I sat near the playpen, waiting for another visitor who had unexpectedly telephoned. This was Dr. Austin H. Riesen, an eye specialist who had heard of Dr. Shaw's patient and wanted very much to see Jayme for himself. Since he was vacationing near Tucson, he intended to drive over, and Jo had given him careful directions, taking a long time over it so that he would not lose himself in a maze of farms and ranches.

"Come on, you klutz," said Jo lovingly to Baba, who is not very bright. "If you could flap your ears, you'd fly. These two are half brother and sister, you know—Danny fathered them both. No, don't wrestle too hard with Jayme, Hahshani; you know better than that."

The three chimps sucked at their bottles. A door slammed, and Paul came in, singing cheerfully, "Happy birthday to you, happy birthday to you, happy birthday, dear Jayme. I'd better get her out of there for the feeding." He did so, first spreading a blanket on the floor. The other two children, having got bored with their meal, scrambled around on the rugs of the playpen. They could easily have crawled or even jumped out, and I asked if they ever did.

"No," said Paul, "not yet. If by accident one of them does bump out onto the floor, he screams the house down. Scared. The day they start getting out is the day they go out to join the big chimps; they're ready."

"And it can't come too soon for me," said Jo feelingly.

Paul went and fetched Aki, and his feeding began. He said, "You know, of course, what the African natives say? That chimps really can talk, but they don't want to because then they'd have to work."

I replied, "I never heard that, but when I was in Central Africa, we would sometimes hear the wild ones screaming, and the Africans said they were dancing out there, and when one of them got the wrong steps, they all beat up on him or her. That's why they screamed."

Aki, whose gums were badly inflamed, drank little of his bottle. We sat there waiting and watching Jayme as she struggled around on her blanket.

"When she used to have her nine o'clock feeding and was just a baby," said Jo, "Paul would sometimes bring her to me in bed with the bottle, and when she'd finished, she just rolled over—this was before the operation, of course—and put her arms around my neck, as if she had to feel a body somewhere near. Whenever we touched her, we had to prepare her first, leaning over and saying her name so as not to startle her. It was a problem protecting her and yet letting her be a chimp. We would sit here and flinch—we still do—when the others bump her or walk over her, yet she's got to have that experience, that company. It ought to go better now."

"Yes, when you teach them, you have to expose them to life," said Paul. "Let them bump her, let her learn to get out of the way."

"She has a sweet face," I said.

Jo said, "Hasn't she? And, I think, a look of wonder. She's had that ever since she came out of the operation. And she's growing so fast; when you feed her, you feel the strength coming, and the determination."

We began to wonder if Dr. Riesen had lost himself. To fill in the time, Jo told about famous breaks for freedom made by some of the chimps. Once a female got out and stayed out for two hours, during Paul's absence. She walked into the kitchen hand in hand with Jo, had a glass of iced tea under

Jo's impotent gaze (Jo didn't want to test strengths with her), then calmly walked out again.

"All of a sudden she went around the house, went back to her cage, marched into it, and sat down. Of course, I rushed to lock her gate. An hour later Paul came home. It was as if somehow or other she knew he was coming. Sometime later a girl was visiting me for a half hour or so, I forget why, and I told that story, and she looked out the window and said, 'Isn't that one out now?' And I looked out and said, 'Yes.' I recognized her—a female from what we call the baby room—and I knew I couldn't get her back myself, so I tried to reach Paul at one of the groceries where he gets supplies. Luckily I just caught him in time, and he said he would come right away—because if one was out, lots of them could get out the same way. I knew the drive from Tempe takes forty-five minutes; somehow we'd have to manage before he got home. Then I saw another one I recognized, and he wasn't from the baby room. That blew my mind. It meant two rooms were open at least, letting out big animals I couldn't manage at all. I called a student who was helping out that day and told him to lock up the baby room. The escaped young ones were trying to get into the office, and they succeeded. None of them are happy outside, you know, after a little while. Meantime, all I could do was stand at the window and watch what was going on. Geronimo was there with some others, and they went back to the cages and let out more. I was in terror because of the dogs, especially the two little ones; the big ones could take care of themselves. Two more big female chimps came out and started for the gate. I didn't recognize them; if I'd known them, I could have gone out and at least herded them back toward the house, but I just waited. The two females climbed the fence and saw the water in the canal; one came rushing back, but the other went on. Then four young ones appeared. What was I to do? At that moment Paul came roaring in in the van, having done the trip in thirty minutes. It's impossible, but he did it. In the meantime, the coyote was howling and howling. Paul walked down to the chimp house, turned the corner, and met one just sitting there calmly, waiting for him to come and let her in. He took another big one by the hand; she scratched him but came in without any further struggle. He rounded up the big ones, leaving four little ones still out. Paul told the students to come into the office and keep quiet. Then he went out of doors, clapped his hands, and shouted, "Okay, everybody. Fun's over; let's go back in!" And everybody gathered around him and fell in behind Paul, and he marched them in like the Pied Piper, and I could have *killed* him. He might at least have made it look harder."

I asked how they had got out in the first place, and Paul said, "The big ones broke their lock on the cage; then they went and let out the little ones. We have a better system now."

"There were still two in the baby room who could have come out and didn't," said Jo. "You ought to have seen the bathroom afterward. They'd mixed up deodorant, perfume, Kitty Litter, two wigs I was using because I'd lost my hair after an anxiety crisis—it was all together in the toilet, indescribable. They hadn't broken anything, just mixed it. I washed and washed the wigs afterward, but somehow I couldn't bring myself to use them, so now the chimps have them as toys. Anyway, my hair grew back."

Still no Dr. Riesen, so we talked about what chimps eat besides yogurt, vegetables, and fruit. "Steaks," said Paul. "I've seen chimps eat whole steaks. They don't have to have meat, of course, if they get enough protein; the steaks are a matter of adaptation, like with people. If they can steal things, they'll eat them. I've known a chimp to eat liver and onions. And, of course, spaghetti. I knew a tiger that loved watermelons; in fact, I've never known a tiger who didn't like watermelon. In order to feed a lion properly, you should at least once a week wrap the meat around some vegetables. Once a week you cover the meat with cod-liver oil. Or if your tiger doesn't get the chance to eat bones, you cover his meat with bones about once a week. At the Berlin Zoo we had a butcher who made a great point of saving the entrails of the animals he butchered for our carnivores. Wonderful for them."

Then the doctor arrived to see Jayme. He said she was in wonderful condition and her eye was doing fine.

Back in New York, I telephoned Jo to check up on a couple of points and asked how everybody was.

"Fine," said Jo. "Aki's cut that tooth at last. Oh, and we have a new baby, a female; cutest thing you ever saw."

"Is her mother raising her?"

"So far," said Jo. "I'm keeping my fingers crossed."

$\mathcal{A}$FTERWORD

THE READER MUST HAVE NOTICED that Jane Goodall is not a main character in this book, though inevitably she plays an important part. Decidedly, I didn't want to leave her out, but the fact is she has told us everything about herself, or almost everything, and repeating it would be presumptuous.

Jane Goodall's second husband, Derek Brycesen, died a few years ago. One of those British-born residents of an African country who chose to take on the nationality of their homeland when it became independent, he was director of national parks in Tanzania. In that capacity, of course, he was able to help his wife immeasurably. For instance, it was their joint decision to change the Gombe Stream Reserve into a Tanzanian project, training native Tanzanians rather than outsiders in her work. This was a move that effectively canceled any plans for future occurrences like that one notorious case of the American students at Gombe who were kidnapped and held for ransom. Since his death Ms. Goodall has devoted part of each year to collecting money in order to set up a scholarship in his name.

"Roughly, I spend a third of the year at home in England when my son has his holidays from school," she said, "and a third lecturing, collecting money to set up the scholarship at Cornell, in Brycesen's name. The other third is spent in Africa."

What of the other well-known visitor to Africa, Dian Fossey of the mountain gorillas? There again I hesitated to add to the record because Ms. Fossey published her book *Gorillas in the Mist* in 1983. Yes, she, too, was advised by Louis Leakey and used the technique he devised, living near the apes and watching them until they lost their fear of her.

I had also intended to write at length of Ada Watterson Yerkes, the wife of Dr. Robert Yerkes, who with her husband wrote the standard book *The Great Apes*. At a time when few women went in for science, she became a botanist. But after marriage her life inevitably took her into contact with animals rather than plants. She had to mother a number of young apes, notably the famous Chim and Panzee, Yerkes's first specimens, but it is probable that her most remarkable experience in primatology was in 1915 when a former student of her husband, Dr. Gilbert van Tassel Hamilton, invited the whole family out to California. Hamilton, a psychiatrist, had a private patient there who kept several monkeys and a young orangutan. It was especially tempting for Yerkes to observe the latter at his leisure. Because of the war, he was unable to carry out a plan he cherished, to visit Wolfgang Köhler in the Canaries. Köhler, author of *The Mentality of Apes*, was working there with a community of chimpanzees. An orangutan, however, was then even rarer than a chimpanzee, and Yerkes was eager to observe this one. So it came about that Ada Yerkes, who had two small children (ages two and four) to take care of, also had to look after the exotic young ape. Her bibliography gives an idea of how divided her interests must have been.

" 'Modifiability of Behavior In *Hyroides Dianthus* V.' *Journ. Comp. Neurol. and Psychol.* 1906" and " 'Mind in Plants.' *Atlantic Monthly.* November, 1904" contrast vividly with the title of a paper in the *American Museum Journal* for 1917, by Robert M. Yerkes and Ada W. Yerkes, "Individuality, Temperament and Genius in Animals." By the same authors we also have Chapter 17 in *The Problem of Mental Disorder*, entitled "The Comparative Psychopathology of Infrahuman Primates" (New York: McGraw-Hill, 1934). There is, of course, *The Great Apes* as well, and various papers. In spite of these, however, Mrs. Yerkes's interests were not in primatology except at second hand, and at last, though reluctantly, I gave up the idea of including her in this account of women with apes.

There remains one more remarkable woman, Nadia, or Nadezhda, Ladygin-Kohts, who crops up in Dr. Yerkes's book *Almost Human*, and occupies a considerable part of his correspondence with women who kept apes. Against great odds she made an important contribution to primatology. Her letters to Yerkes begin in November 1924, when she wrote to him from the Zoological Laboratory of the Darwinian Museum in Moscow. Her husband, Aleksandr, she explained, was curator of the museum, an important place, we can assume, for Russian intellectuals, who wholeheartedly accepted the theory of evolution. Mme. Kohts introduced herself as directress of the laboratory and lecturer of comparative psychology at the state university in Moscow. She announced that she was sending Dr. Yerkes a copy of some studies she had made of a chimpanzee. She had managed, she said, to avoid in her research the Clever Hans Syndrome, this being the name for a trap feared by wary animal behaviorists the world over. It is named after the so-called calculating wonder horse of Berlin, who gave answers to mathematical—or, indeed, any—questions posed by humans by watching for unconscious clues on the part of his questioner. Translated, the title of Mrs. Kohts's paper is "Animal Behavior and Visual Discrimination in Young Chimpanzees." The author had for some time owned a young chimp named Jony (or Ioni, or Joni) and worked with him, producing proof that the ape could tell colors apart and see them as they are seen by us human primates. It was not known until then that all primates—monkeys, apes, and people—possess this faculty in common, and Mrs. Kohts's was a vitally important piece of work. She achieved her results mainly by inducing Jony to match cloth samples, red to red and so on.

A colleague of Yerkes, the Russian-born Dr. Alexander Petrunkevitch, translated her paper for Yerkes, who was much excited by her accomplishment and wrote to tell her so, congratulating her on having done so well in spite of what must have been discouraging conditions in the Russia of the early twenties. He said that he and Petrunkevitch hoped that the rest of her work—for she said she had produced only half of it—could be published and that they were trying to get this done in America. He asked her a lot of questions. What was the age of her chimpanzee when she acquired him? Did she know anything of his background history? How old did she think he was when she first began to experiment with him? What was his attitude toward experimental work at present? In one place she had referred to the chimp as "weeping"; had she in fact ever seen him shed tears? Had he ever tried to construct a nest in a tree? Was she continuing to work with Jony, and had she managed to get other chimps? He said he was sending her copies of his own publications, "which might come in useful." A few months later he wrote telling her that he and Petrunkevitch had written a digest of her paper and sent it to the *Journal of Comparative Psychology*, which would probably publish it in the spring, and he would like, if possible, to obtain some of the illustrations she had used.

Mrs. Kohts responded as carefully as she could to all the eager questions.

She thought that Jony, now dead, had been about three years old when she got him through the official animal purveyor, who had—of course—bought him from a sailor. Jony had died two years later. During the time she owned him she had made many photographs of his expressions and attitudes and kept a diary which she meant to use in later writings. No, there was never a sign of tears in his eyes, though he would cry and scream for hours if left alone; he was always very eager for human company. Such moisture as could be seen in his eyes was due to colds, of which he had many. No, he had not tried to build nests; he was terrified of trees, as a matter of fact, and always gripped her tightly when she took him for walks in the woods. But when he went to bed, he liked to take things with him, twigs, leaves, and frayed bits of cloth, which he tried to put together somehow, and she thought these actions indicated rudiments of nest building. Unfortunately at the time Jony died Russia was, even more than usual, out of touch with the rest of the world, and Mrs. Kohts had been unable to get more chimpanzees to further her research.

Though correspondence was evidently quite possible at this time between Russia and the States, even now, according to Mrs. Kohts, it was very difficult to get books from abroad; for example, she had none of Yerkes's works. He now sent her papers, however, and she seems to have received them. These cordial exchanges were interrupted for some weeks, and when the silence was at last broken, Mrs. Kohts explained it by saying that she had been ill and then had had to nurse her "firstborn" child, who had been ill in his turn. Everybody was now recovered, and she had returned to her research. She was very grateful for all the praise she was receiving in America, and her study of chimpanzee reactions and feelings was under way again. She hoped it would not be too long before it was published. It delighted her to hear, she said, that Dr. Yerkes's work on "chimpanzee intelligence," studies of the vocalizations of chimps which he produced with a musician, Blanche Learned, who interpreted and annotated the sounds, was soon to appear. Late in 1925 Yerkes told Mrs. Kohts that they now had a department of psychology at Yale and were able to keep several chimpanzees in their laboratory for observation. Besides, Köhler was visiting the university.

The correspondence continued. In February 1928, Mrs. Kohts wrote that she had prepared "a little surprise": She had dedicated her new book on the behavior of the rhesus macaque to Dr. Yerkes. In his thank-you letter, he said he had just come back from a session in Sarasota with a circus gorilla, Congo. But where, he asked somewhat sternly, was the publication of her remaining material on Jony? Mrs. Kohts had to reply that the rhesus material, though it was indeed coming out at last, had been delayed because of a paper shortage. (Probably the same kind of difficulty faced her attempts to publish the chimpanzee material.) However, she said, it was now possible to get books from abroad through an approved agency in Germany, though naturally this took a long time.

There was, of course, little actual back-and-forth visiting between the countries, but at least twice friends of Yerkes's from America, visiting Russia, made contact with the Kohts couple and brought back messages. Aleksandr and Nadia Kohts urgently wanted a good photograph of Yerkes, at least one, because they were having a bust made of him for their museum. Finally, in August 1928, Yerkes wrote to one of these friends to say that he was considering a visit to Moscow and a number of other Continental science centers. Within a year or two, he said, he hoped to arrange it, and he asked what his friend thought such a trip would cost—a three months' visit including Moscow, Leningrad, various Swiss and German towns, Brussels, England, and Edinburgh. It did not take him a year to put the plan into effect; in January 1929 he was writing to the Kohtses to announce that he was thinking of that journey and had now added Africa to the itinerary. Would he be likely to find them at home, he asked, if he arrived about the middle of June? He added that he would probably be accompanied by his daughter, Roberta, who had just been graduated from college. He said he would be very grateful if they could advise him on practical matters, such as where to stay in Moscow. He seems to have sent the photograph at the same time as the letter, because Mrs. Kohts's joyful reply (dated April 10) mentions having got it. What she really wanted to say, however, was that she and her husband were excited and happy at the thought of the approaching visit. They would do their best, she said, to make sure that the visitors would have a good time, and she ventured to suggest that they stay at the museum, with the Kohts family. They could be accommodated in two rooms that Mrs. Kohts herself used for a study but hardly ever entered during the summer vacation. The accommodation and food might seem poor to Americans compared with what they were used to, but at least they could stay at the museum while they looked for better lodgings—perhaps in one of the hotels? She named several of these but insisted that she herself, and her husband, would be very happy to have them. They would be delighted to take their guests sight-seeing in Moscow. Mrs. Kohts's faulty English, as she called it, would be compensated for by taking along with the party some of their young friends who were better linguists than she. What other languages could Dr. Yerkes and his daughter speak? Her husband spoke perfect German, but she herself spoke equally badly English, French, and German.

Late in May Dr. Yerkes was able to announce that he and Roberta would arrive in Moscow on or about June 15 and that they certainly accepted Mrs. Kohts's kind invitation to stay with her family. It would be for only a few days at any rate, as he planned to move on to Leningrad to see Professor Pavlov and some other scientists. He could not promise to be fluent in any language but English, as the German he used to know had all been forgotten, but Roberta was good in French and had a rudimentary knowledge of German.

"I don't really like languages at all," he confessed.

Naturally we cannot follow the visit because correspondence ceased for the summer, but through mentions in later letters we know that Dr. Yerkes was told of Mrs. Kohts's latest work, which was to observe and note down the behavior of her small son and compare it with the diary and notes she had kept of Jony. Did this work, as well as her remaining material on Jony, ever get published? We do not know, though Roberta Blanshard remembers that the visit was delightful.

"They made us as comfortable as possible," she said, "and this was at a time, remember, when they might well have been in trouble for entertaining Americans. We never forgot them. They were sweet."

Only the Russians can tell us if Mrs. Kohts's later work was ever published. The child, if he survived, would be about fifty-eight now. Mrs. Blanshard does not know; I don't know. Can anybody help us out?

BIBLIOGRAPHY

Benchley, Belle. *My Life in a Man-Made Jungle!* London: Faber & Faber, 1942.

—— *My Friends, the Apes.* Boston: Little, Brown, 1944.

Brewer, Stella Margaret. *The Chimps of Mount Asserik.* New York: Alfred A. Knopf, 1978.

Harrisson, Barbara. "The Nesting Behaviour of Semi-Wild Juvenile Orang-utans." Reprinted from the *Sarawak Museum Journal*, Vol. XVII, Nos. 34–35, 1969.

——*Orang-utan.* New York: Doubleday, 1963.

Hoyt, Augusta. *Toto and I.* Philadelphia and New York: Lippincott, 1941.

Linden, Eugene. *Silent Partners: The Legacy of the Ape Language Experiments.* New York: Times Books, 1986.

Lintz, Gertrude Davies. *Animals Are My Hobby.* McBride, 1942.

——*Apes On Stage.* Publishing information unavailable.

Noell, Mae. *Gorilla Show.* Tarpon Springs, Fla.: Noell's Ark Publishers, 1979.

Patterson, Francine, with Eugene Linden. *The Education of Koko.* New York: Holt, Rinehart & Winston, 1981.

Yerkes, Robert Mearns. *Almost Human.* New York: Century Co., 1925.

The author gratefully acknowledges permission to reprint the following:
Excerpts from My Life in a Man-Made Jungle! by Belle Benchley,
© 1942 by Faber & Faber Ltd. Reprinted by permission of the publisher.

Excerpts from The Education of Koko by Francine Patterson and Eugene Linden, © 1981 by Henry Holt and Company, Inc. Reprinted by permission of the publisher.

The photographs in this book appear courtesy of the following: BELLE BENCHLEY: © Zoological Society of San Diego; QUINTA PALATINO, MADAM ABREU: Robert Yerkes papers, Yale University Library; ROBERT YERKES WITH CHIM AND PANZEE: Yerkes Regional Primate Center of Emory University; ANDRÉS, PRIMATE QUARTERS AT QUINTA PALATINO, PORTRAIT OF CUCUSA AND ANUMÁ: Yerkes Papers, Yale University Library; TOTO, AUGUSTA HOYT, AND TOMÁS originally appeared in Toto and I by Augusta Hoyt (publication information in Bibliography, page 179); GERTRUDE LINTZ AND CHIMPS originally appeared in Animals Are My Hobby by Gertrude Lintz (publication information in Bibliography, page 179); "FEROCIOUS" GARGANTUA: AP/Wide World; "PLAYFUL" GARGANTUA: Alfred Eisenstaedt/Life; MAE AND BOB NOELL AND THE GORILLA SHOW: collection of Mae Noell; BOB NOELL AND TOMMY SIDE BY SIDE: Norman Zeisloft; STELLA BREWER, RAPHAELLA, AND WILLIAM originally appeared in The Forest Dwellers by Stella Brewer (publication information in Bibliography, page 179); BARBARA AND TOM HARRISSON, BARBARA IN WATER: © Larry Burrows collection; BARBARA HARRISSON: collection of Barbara Harrisson; JO FRITZ: collection of Jo Fritz; NADIA KOHTS WITH YERKES AND TOLSTOY: Yerkes Papers, Yale University Library; NADIA KOHTS: Yerkes Regional Primate Center of Emory University. PHOTO RESEARCH BY ESSEN ILI GÖKNAR.

ABOUT THE AUTHOR

Emily Hahn, a *New Yorker* staff writer who has herself been an "Eve" to several apes, is the author of more than fifty books of fiction and non-fiction. She is a member of the American Academy of Arts and Sciences. Ms. Hahn lives in New York City.